BIOCHEMISTRY RESEARCH TRENDS

TEMPERATURE-DEPENDENT BIOLOGY AND PHYSIOLOGY OF REDUVIIDS

BIOCHEMISTRY RESEARCH TRENDS

Additional books in this series can be found on Nova's website under the Series tab.

Additional E-books in this series can be found on Nova's website under the E-books tab.

TEMPERATURE-DEPENDENT BIOLOGY AND PHYSIOLOGY OF REDUVIIDS

K. SAHAYARAJ

AND

S. SUJATHA

Nova Science Publishers, Inc.

New York

Library of Congress Cataloging-in-Publication Data

Sahayaraj, K.
 Temperature-dependent biology and physiology reduviids / K. Sahayaraj and S. Sujatha.
   p. cm.
 Includes bibliographical references and index.
 ISBN 978-1-61209-940-8 (hardcover)
 1. Assassin bugs. 2. Assassin bugs--Physiology. 3. Assassin bugs--Biological control. I. Cujata, Es. II. Title.
   QL523.R4S25 2011
   595.7'54--dc22
                   2011008483

Published by Nova Science Publishers, Inc. † New York

CONTENTS

PREFACE

Biological control is one of the most important components of Biointensive Integrated Pest Management (BIPM) strategy, in which natural enemies, namely predators, parasitoids and pathogens, play an important role. Reduviids are reported to suppress more than most of the lepidopteran larvae, few coleopteran, hemipteran, isopteran, and orthopteran pests both in laboratory and field situations. Reduviids have played a major role in suppressing the pest population worldwide and they can be utilized as a biological control agent against a variety of pests. Unfortunately, the biological potential of the reduviids has not been investigated, as large-scale release studies have not been carried out so far. Hence, there appears a better scope for utilizing the reduviids in the biological control programme. The efficiency of natural enemies is affected by environmental conditions, mainly temperature. Scanty information is available on the impact of cyclic conditions of temperature and humidity on the development and physiology of entomophagous reduviid predator, though the impact of temperature on Pentatomidae, Anthocoridae and Miridae were studied in detail. Knowledge of the biology, predatory rate and physiology of two reduviids namely, *Rhynocoris marginatus* (Fab.) and *Rhynocoris fuscipes* (Fab.) over a constant temperature range of 5°C to 35°, the most suited for these reduviids' development, will help in predicting the synchronization of the occurrence of the predator and prey and the effect of the predator on the population density of these predators in the field. Unlike the handful of other texts on this subject, this book is not a compendium of temperature-dependent impacts on predatory insects. Instead, it is an effort to explain and bring out the information about the effect of temperatures on the development and reproduction, bioefficacy and physiology of reduviids in general and two reduviids in particular.

The book entitled *"Temperature-dependent Biology and Physiology of Reduviids"* explains biology (nymphal development, weight gain and survival rate; sex ratio, adult longevity, fecundity and hatchability, morphogenesis), linear developmental model (upper and lower temperature thresholds), biological control efficiency (stage preference, predatory acts and predatory rate). The book draws heavily from insect physiology because the base of knowledge of these fields is highly advanced in developing a base of understanding of virtually every aspect of physiology- as egg and total body macromolecular profile (total carbohydrate, protein and lipids), enzymes (invertase, amylase, protease, esterase), gut bacterial population (species composition and hydrolytic extra cellular enzyme activity), gut protein profile by SDS-PAGE, DNA amplification and genetic similarity in relation to six universal primers and interaction of prey-protein and predator response by ELISA. The book consists of a preface and contents, followed by seven chapters and references, species and subject index. We most hope this book will be a bridge for predator rearing specialists and their stakeholders to use various temperatures under storage conditions to better manipulate in ways that benefit temperature and our environment.

ABOUT THE AUTHORS

K. Sahayaraj, Ph.D., is Associate Professor and Director of Crop Protection Research Centre, St. Xavier's College, an educational Institution dedicated to higher education and research. Recently, he has been publishing an international journal, namely Journal of Biopesticides (ISSN 0976-0341X). Dr. Sahayaraj earned B.Sc. and M.Sc. degrees in Zoology (1985, 1987) and M.Phil. and Ph.D. in Zoology with a specialization in Agriculture Entomology (1988, 1992) from Madurai Kamaraj University, Madurai, India and a PGDCA from the same University during 2001. He also earned PG Diploma in Plant Protection (PGDPP) (2007) from the Annamali University, India. He has been teaching Zoology including Entomology to undergraduate and post-graduate students for more than 17 years. Dr. Sahayaraj has published three books and one proceeding on ecofreindly insect pest management. He has more than 135 publications, including original research papers, book chapters, and popular articles in insect ecology, behavior, biology and physiology, as well as numerous papers on biological control efficacy of reduviids. He has been honored with several awards from regional, national and international agencies.

Over the past 23 years, Dr. Sahayaraj's research efforts have been dedicated to multidisciplinary, integrated approaches to understanding how reduviids distribute and diversify in various ecosystems, and how their adaptive characters can be applied to pest management, especially through bio-intensive pest management. He has operated nine research projects funded by national (DST, DBT, CSIR) and international (IFS) funding agencies, and now he has been operating three projects funded by CSIR, MoEs and MEFs. Dr. Sahayaraj has guided more than 12 Ph.D. students and supervised 8 scholars. This book is an effort of Dr. Sahayaraj and his student Dr. S. Sujatha that a thorough, comprehensive understanding of the basic and applied science behind reduviid predator and temperature interaction can be applied profitably to solve even the most challenging practical problem while storing the reduviids for augmentatives release programme. His efforts have been devoted to increasing our understanding of the basic tenets of the storage of insects in different temperatures, and to apply this basic knowledge to practical aspects of biological -based pest management systems.

Dr. S. Sujatha has completed her M.Sc. and M.Phil. in Zoology (2002, 2003) and Ph.D. in Applied Entomology (2008) from Manonmanium Sundaranar University, India. She has more than 8 publications in reputed journals on Ecofriendly pest management. For the past two years she has been teaching post-graduate students and also guiding for their theses at International Centre for Bioresources Management, Malankara Catholic College, Mariagiri, Kaliakkavilai – 629153, TamilNadu, India

Chapter 1

REDUVIIDS AND PEST MANAGEMENT

Biointensive Integrated Pest Management (BIPM) has been enshrined as a cardinal principle of plant protection in overall crop protection programmes in order to minimize the indiscriminate and injudicious use of chemical pesticides. Biointensive Integrated Pest Management includes factors like natural enemies, botanicals, microbial insecticides, etc. Reducing crop losses, minimizing pesticide use, avoiding pesticide residues, increasing farmers' incomes and enhancing environmental health are the hallmarks of sustainable agriculture. BIPM is more compatible with the environmental components than synthetic pesticides.

Heteroptera represents an important part of the total insect fauna in many agroecosystems on the basis of their numbers rather than biomass. Most heteropterans are smaller in size and exhibit food regimes from strict phytophagy to zoophagy. They are rarely host-specific but often show clear preferences for definite strata. There are many species of "true bugs" (Order Hemiptera), such as tarnished plant bug, which feed on plants; however, a number of them are also predators of pest species. The terrestrial bugs most often encountered in a crop field are the "assassin bugs" or the reduviids (Family Reduviidae). Reduviids are particularly numerous and diverse on agricultural crops but most species' complexes are identical on semi-arid zones and scrub jungles bordering agroecosystems. Although a cultivated agroecosystem does not appear to be favoured by Heteroptera, it becomes attractive during the flowering season. Reduviids are more resistant to chemical sprays than the coccinellids and spiders. Their adjustability to ecological factors and to the secondary effects of phytosanitary sprays are the features that make Heteropterans, especially Reduviids potentially good biological control agents. These voracious predators are abundant, cosmopolitan and larger in size than other predaceous land bugs, which make them to be more effective biological control agents.

1.1. REDUVIID PREDATORS

Reduviid (Hemiptera: Reduvidae) predators are the largest terrestrial bugs, globally comprised of 6250 species and sub-species, 913 genera and 25 subfamilies (Maldonado, 1990). According to Ambrose (2006), in India, 464 reduviid species belong to 144 genera and 14 subfamilies were present. These bugs are particularly numerous and diverse on agricultural crops but most species' complexes are identical on semi-arid zones and scrub jungles bordering agroecosystems. Reduviids are more resistant to chemical sprays than other potential biological control agents. Their adjustability to ecological factors and to the secondary effects of phytosanitary sprays are the features that make Heteropterans, especially Reduviids potentially good biological control agents. These voracious predators are abundant, cosmopolitan and larger in size than other predaceous land bugs, which make them to be more effective biological control agents.

1.2. DISTRIBUTION AND DIVERSITY

Reduviid predators have been recorded from many agroecosystems such as sugarbeet (Ehler et al., 1997), tropical legume sesbania (Sileshi et al., 2001), oilseed brassica (Ruberson and Williams, 2000), oil palm (Cheong et al., 2010), soybean (Grundy, and Maelzer, 2000), cotton (Shower and Greenberg, 2003; Grundy, 2007), lady's finger, chilli, cinnamon, citrus, coconut, cowpea, groundnut, maize (Altieri et al., 1978), mango, mustard, pigeonpea (Minja et al., 1999; Ambrose and Claver, 2001), potato, pumpkin, rice, sugarcane, sunflower, sweet potato, tobacco (Marques *et al.,* 2006), teak (Das and Ambrose, 2008a and 2008b), wheat (Sahayaraj, 2007a), cocoa (Louis, 1974) and apple orchard (Sackett *et al.,* 2007).

In addition to the agroecosystems, it also dwelled in the storage places like warehouses, goudowns, grain mills and shiftment areas. For example, *Amphibolus venator* (Klug) (Hemiptera: Reduviidae) is a predator of many stored product insects. It preys on *Trogoderma granarium* (Everts), *Tribolium castaneum* (Herbst), *Corcyra cephalonica* (Stainton), *Latheticus oryzae* (Waterhouse) and *Alphitobius diaperinus* (Panzer) (Pingale, 1954; Haines, 1991). In a warehouse trial, Pingale (1954) indicated that *A. venator* is effective for the control of *Ephestia cautella* (Walker) and *A.diaperinus*. *A. venator* has frequently been found in shipments of groundnuts from Africa to England (Hill, 1990) and also in warehouses in Thailand. It appears that *A. venator* adapts to the habitat of rice milling facilities. Another assassin bug, *Peregrinator biannulipes* (Montrouzier and Signoret), is well known as a natural enemy of stored product pests. It preys on moths, as well as *Anagasta kuehniella* (Zeller), *Plodia interpunctella* (Hübner), *C. cephalonica*, *Pyraris farinalis* (Linnaeus) and beetles, *T. castaneum*, *T. confusum*, *Stegobium paniceum* (Linnaeus) and *Lasioderma serricorne* (Fabricius) (Tawfik et al., 1983; Awadallah and Afifi, 1990). *P. biannulipes* was also found in the same rice milling facility as *A. venator*.

Each agroecosystem has its own micro climatic condtions, which is governed by the temperatures, relative humidity, wind velocity, etc. However, very little has been published on the effect of temperature on development, survivorship, biological control potential and physiology of reduviid predators. Hence it is imperative to know the impact of temperatures against the life stages of these unexplored important natural enemies. We undertook an effort to find out the impact of various constant temperatures on the biology and physiology of two important reduviids from India. We also highlighted the published literature available in the same line.

1.3. WHY REDUVIIDS COULD BE CONSIDERED IN PEST MANAGEMENT

Reduviids are well known as generalist predators feeding on a variety of preys. These predators are considered as dominant ones over coccinellids (Sahayaraj and Raju, 2006), *Chrysoperla carnea* (Rosenheim, 1998), spiders (Wignall and Taylor, 2008; 2009), and as also having a less predatory risk (Gwynne, 1989). Furthermore, reduviid predators are resistant/non-toxic against both synthetic (Sahayaraj, 1991; George and Ambrose, 1998; Grundy et al., 2000; James and Vogele, 2001; Zulkefli et al., 2004) and biopesticides (Jaronski et al., 1998; Fadare and Osisanya, 1998; Fadare and Amusa, 2003; Namasivayam, 2005; Sahayaraj, 2007a; Sahayarj et al., 2002, 2008). Moreover, reduviids are much larger in size than other Hemipteran predators like *Nabis* (Nabidae), *Geocoris* (Lygaeidae), *Orius* (Anthacoridae), *Lygus* (Miridae) and *Podisus* (Pentatomidae) and they successfully attack and consume larger preys.

The subfamily Harpactorinae is the largest and most poorly studied subfamily of Reduviidae. Among various reduviid predators reported in India, the predominant and widely distributed reduviids were *Rhynocoris marginatus* (Fab.) and *Rhynocoris fuscipes* (Fab.) (Harpactorinae). Further, studies have shown that these two predators were wonderful, voracious hunter reduviids, which can be utilized in pest management programmes. Hence it is worthwhile to study their biology and physiology in relation to various constant temperatures. In the forthcoming sessions, we have highlighted the importance of these two reduviids throught the available publications.

1.4. 1. Rhynocoris marginatus (Fab.)

Rhynocoris marginatus (Fab.) is considered to be the most important predator of many pests. All the five nymphal instars and also the adults were obligatory potential predators of many economically important crop pests. It has been previously reported that under laboratory conditions, the temperature is indispensable to the eclosion and moulting of the reduviids, which ranged either between 16 °C and 34°C (Gomez - Nunez and Fernandez, 1963) or 15°C and 35°C (Okasha, 1964, 1968a, b and Jones and Sterling, 1979).

Population dynamics of the reduviid predators in relation to various biotic and abiotic factors have been reported by Ambrose (1980), Vennison (1988), Sahayaraj (1991), Kumaraswamy (1991), Sahayaraj and Ambrose (1993), and Das (1996). The population density of a particular reduviid depends upon the biotic and abiotic factors (Goel, 1978; Haridass, 1987; Ambrose and Livingstone, 1989; Vennison and Ambrose, 1990; Sahayaraj 1991; Kumaraswami and Ambrose, 1993 and 1994; Ambrose and Maran, 1999a). Recently, Dhanasing and Ambrose (2006) reported the seasonality of the reduviid predator's population in Thoothukudi District, Tamil Nadu, India. The climatic factors indirectly governed the distribution and density of the assassin bugs in any natural ecosystem as observed by Ambrose and Rani (1991), Ambrose and Rajan (1995) and George and Ambrose (2000b). Sahayaraj (2007a) reported that the reduviid population has been observed abundant in dry regions, even at lower rainfall and relative humidity and moderate temperature.

Sahayaraj (2007a) reported *Rhynocoris marginatus* as one of the largest and most easily recognized assassin bugs found in semi-arid, scrub jungle, tropical forests and agroecosystems. The population of this bug is prevalent from January to June and from November to February. However, peak population was observed during March and April. In Tirunelveli district, *Rhynocoris marginatus* was recorded both in *rabi* (September to January), and *kharif* (June to August) seasons. It was also recorded that distribution of the reduviid predators was influenced by wind velocity. During the same period, Sahayaraj and Raju (2004) reported the occurrence of 11 hemipteran, nine coleopteran, four lepidoptera, two each of dermapteran and orthopteran and one isopteran species and among them, 12 species were present both in *kharif* and summer seasons, whereas others were present exclusively during *kharif* season. They also reported natural enemies belonging to Coleoptera (6 species), Araneidae (5 species), Hymenoptera (6 species), Reduviidae (3 species) and Odonata (1 species) from both Thoothukudi and Tirunelveli districts of Tamil Nadu, India.

In 2004, Sahayaraj and Raju (2004) took a survey of reduviids in Kanyakumari, Madurai, Tirunelveli, Thoothukudi and Virudunager districts of Tamil Nadu. Results showed that Harpactorinae had the maximum number of species followed by Peiratinae and Reduviidae and Stenopodainae. In total, nine species under four sub-families were recorded from 17 agricultural zones of five districts of Tamil Nadu. Among the nine reduviids, population of *Rhynocoris marginatus* was maximum followed by *Rhynocoris fuscipes* (Fab.), *Rhynocoris longifrons* (Stal), *Oncocephalus annulipes* (Stal), *Acanthaspis quinquespinosa* (Stal), *Ectomocoris cordiger* Stal, *Ectomocoris tibialis* (Distant), *Rhynocoris kumarii* (Livingston and Ambrose) and *Acanthaspis pedestris* Stal. These findings suggest that distribution of reduviids in groundnut is well recorded and initiative has been taken to find out their role in pest management. Further investigations were undertaken to study the bionomics and biological control potential of *Rhynocoris marginatus* with reference to groundnut pests.

Rhynocoris marginatus adults are usually 18 to 21 mm long and around 160 mg in weight. It is a bright coloured predator and exists in three different morphs (Ambrose and Livingston, 1988), distinguished by the colours of their connexivium: niger (black), sanguineous (red) and nigrosanguineous (black and red). The head is narrow with a distinct neck behind the often-reddish eyes. The long, curved mouthparts form a beak (rostrum), which is carried beneath the body, with the tip fitting in a groove on the underside of the body. The middle of the abdomen is often widened, so the wings incompletely cover the width of the body. Further, it was recorded that the females are larger than males. Females lay eggs in tight, upright clusters under the surface of the leaves or in the soil. Nymphs resemble the adults, but are smaller, wingless and may be brightly coloured. This bug, when young, feeds on small larvae and nymphs, aphids and mealy bugs and later attack medium and larger caterpillars of Lepidoptera, nymphs and grubs of Hemiptera and Coleoptera, respectively and also feeds on adults of Isoptera and Hymenoptera.

Total nymphal instar period ranges from 39 to 85 days depending upon the prey and abiotic conditions in which they are reared. Female biased sex ratio is common in *Rhynocoris marginatus*. It has five nymphal instars. Females live longer than males and they mate with multiple males. Development and reproduction of reduviids are influenced not only by ecological factors, but also by biotic factors, especially the prey. Females exhibit parental care; they take care of their eggs and also nymphal instars early on. Prey types, the number of preys offered, and the type of oligidic diet provided have influenced biology parameters. Prey species influence the development and reproduction of *R. marginatus*. Our study showed that small preys like *Odontotermes obesus* delayed the developmental period (39 days) and reduced the fecundity (155 eggs/female) and hatchability (29 %). The grasshopper, *Chrotogonus* sp. was found to be the most suitable prey among all the tested prey categories. *Chrotogonus* sp. reduced the incubation period (7 days) and stadial period (46 days), increased the adult longevity (67 and 91 days for male and female, respectively), fecundity (210 eggs/female), hatchability (91 %) and adult size (153 mg and 1.92 cm) of *Rhynocoris marginatus*.

The factors that determine prey suitability for insect predators can be divided into nutritional and non-nutritional factors. Competition, space, and starvation are the crucial factors, which regulate predatory behaviour of this bug (says Ambrose et al., 1985). For a prey species to be suitable, it must provide all nutritionally important factors, such as proteins, carbohydrates, lipids, vitamins and minerals in a balanced proportion and concentration to meet the predator's metabolic demands. Studies were undertaken to know the impact of groundnut pests like *Spodoptera litura* alone and *Helicoverpa armigera* and *Spodoptera litura* in combination on bionomics of *Rhynocoris marginatus*. When three days (17 preys/nymphal stage) and four days (18 preys/nymphal stage) old *Spodoptera litura* were provided as prey; the total nymphal developmental period was 37.66 days. Females lived longer (69 days) than males (47 days). Later instead of providing the very early stage of the host (3 or 4 days), third and fourth instar larvae of *Spodoptera litura* were provided and biology was studied. By consuming 234 third and fourth instar larvae of *Spodoptera litura*, the reduviid completed its nymphal stages within 47 days and produced maximum number of females (0.91:1.0, male:female). This reduviid needed an average of 22.37 preys to complete the nymphal period. It was observed that female bugs lived 42 days more than males. Females started laying eggs within 19 days after their emergence and laid as many as 405 eggs per female.

Since *Helicoverpa armigera* occurs in large numbers and is a voracious feeder causing extensive damage even at low population density in groundnut, its influence on the biology of *Rhynocoris marginatus* was studied along with that of *Spodoptera litura* for comparison. Studies showed that *Helicoverpa armigera* reduced the pre-imaginal period (2 days) and post-oviposition period (9 days) of *Rhynocoris marginatus*. *Helicoverpa armigera* not only reduced the nymphal developmental period but also increased the nymphal weight, enhanced the egg laying period (12 days), and the total number of eggs laid by a female (50 eggs/female). Reason proposed for these enhancements is that *Rhynocoris marginatus* consumed more amounts of carbohydrate (84%), protein (95.7%) and lipid (57.1%) from *Helicoverpa armigera* than from *Spodoptera litura* (73.1, 79.3 and 46.7 % for carbohydrate, protein and lipid, respectively). Studies revealed that groundnut pests support the development and reproduction of the reduviid. Moreover, between the two lepidopteran hosts, *Helicoverpa armigera* supports the quicker development and maximum fecundity of *Rhynocoris marginatus* than *Spodoptera litura*. Hence, for rearing this reduviid in the laboratory, *Helicoverpa armigera* fourth instar larvae can be used as a prey (Sahayaraj et al., 2004).

Rhynocoris marginatus is a predator of insect pests of agricultural as well as forest vegetation. The prey record includes various families of Coleoptera, Hemiptera, Hymenoptera, Isoptera, Lepidoptera and Orthoptera. Laboratory results revealed that *Rhynocoris marginatus* adults consumed the maximum of *Aproaerema modicella* (18.37 larvae/day) followed by *Helicoverpa armigera* (16.93 larvae/day), *Spodoptera litura* (16.12 larvae/day), *Amsacta albistriga* (4.06 preys/day) and *Aphis craccivora* (0.26 adults/day). However, younger nymphal instar predators of *Rhynocoris marginatus* consumed maximum numbers of *Aphis craccivora* (6.47 adults/day). Hence, this predator could be mass reared and released augmentatively; evaluation of its biological control potential under field condition is imperative.

It is a selective biological control agent of many agricultural and forest insect pests like *Earias fraterna* (Fab.) (Ambrose, 1988), *Papilio demoleus* (L.) *Earias vittella* (Fab.) (Nayar *et al.,* 1976), *Corcyra cephalonica* (Stainton) (Bhatnagar *et al.,* 1983), *Helicoverpa armigera* (Hubner) (Ambrose 1987), *Mylabris indica* (Faust), *Mylabris pustulata* (Fab) (Imms, 1985 and Nayar *et al.,* 1976), *Achaea janata* (Linn.), *Oxycarenus hyalinipennis* (Costa) and *Approarema modicella* (Deventer) (Sahayaraj, 1995a, 1995b, 1995d; Sahayaraj *et al.,* 2003), *Spodoptera litura* (Fab.) and *Amsacta albistriga* (Walker) (Sahayaraj, 2000), etc.

Recently Sahayaraj (2007a) narrated the biological control potential of *R. marginatus* on four groundnut pests under laboratory conditions. George *et al.* (2002) observed the nutritional influence of prey on the biological control and biochemistry of this bug. Previously, the impact of space (Vennison, 1988), mating behaviour (Ambrose and Livingstone, 1985), starvation (Ambrose *et al.,*1990a, 1990b) and prey influenced (Ambrose and Claver,1996) the biology of the bug. Bioefficacy and prey size influenced the developmental period (Sahayaraj, 1995a, 1995b,1995c) and predatory potential of the bug (Ambrose and Claver, 1996). Ecotypic (Ambrose, 1987) and polymorphic diversity (Ambrose and Livingstone, 1978; Vennison and Ambrose, 1988) of the reduviid was also documented. There was a lesser effect of biopesticides like Neem Gold and NPV on the incubation period and hatchability of *R. marginatus* eggs (Sahayaraj and Paulraj, 1999). Biocontrol efficiency (Sahayaraj, 1999, Ambrose and Claver, 1998) and possible integration of this bug either alone (Sahayaraj and Martin, 2003; Sahayaraj and Balasubramanian, 2008) or in combination with botanical insecticides (Sahayaraj and Ravi, 2007) has been highlighted. Sahayaraj *et al.* (2003) observed the effect of two biopesticides namely Neem Gold and NPV on the eggs and nymphal instars of the predator. Similarly, Sahayaraj and Martin (2003) witnessed augmented control in groundnut pests. Recently Sahayaraj *et al.* (2007a, b) and Sahayaraj and Balasubramanian (2008) studied the prey influence on salivary gland and gut enzyme profiles of the reduviid. From the above-mentioned informations, it is very clear that *R. marginatus* is a good biological control agent.

1.4.2. Rhynocoris fuscipes (Fab.)

Rhynocoris fuscipes (Fabricius) is a crepuscular, brightly coloured (black and red), entomophagous, harpactorine reduviid found in concealed habitats such as underneath the stones and crevices (Ambrose and Mayamuthu, 1994; Ambrose, 1987; George *et al.*, 2000; Sahayaraj and Selvaraj, 2003). Bioecology of this predator has been studied by Ambrose and Livingstone (1986). Ambrose (1997) first described the mating behaviour of this bug; later, the impact of mating on oviposition pattern and hatchability (Vennison and Ambrose, 1986). Moreover, food requirement has also been known for this predator (Ambrose and Kumaraswami, 1993). Previously, impact of sex, starvation, antennectomy, eye blinding and tibial comb coating on the predatory behaviour of this reduviid was studied (Ambrose and Mayamuthu, 1994).

When it is found in an agroecosystem, it predates upon various insects like *H.armigera* (Hubner), *C. cephaionica* (Stainton), *A.janata* (Linn.), *Plutella xylostella* (L.) *S. litura, Myzus persica* (Sulz.), *Lygus hespes* (Fabricius), *Rhaphid opaipa* (Thunb.), *Foveicollis lucas* (Distant.), *Semiethisa pervolagata* (Walker.) (Singh, 1985), *Epilacrisia stigma* (Muls.) (David and Natrajan, 1989), *Cryptosilla pyranthes* (Linn.) (Hiremath and Thondarya, 1983), *Calocoris angustatus* (Leth.) (Ambrose, 1980); *Patanga succincta* (Linn.), *Dysdercus cingulatus* (Fab.), *Earias vittella* (Fab.) (Singh and Singh, 1987), *Nezara viridula* (Linn.) (Singh and Gangrade, 1975); *Perigrinus maidiis* (Ashm.), (Ponnamma *et al.*, 1919); *Spilosoma obliqua* (Walker.) (Cherian and Kylasam, 1939); *Myllocoris curvicornis* (Fab.), (Cherian and Brahmachari, 1941); *Aulacophosa foveicollis* (Fab.), (Ambrose, 1995); *Pleopidas mathias* (Fab.), *Clavigarata gibbosa* (Spinda.), *Clavigarata horrens (*Distant.), *Dolycoris indicus* (Stal.) (Das, 1996). Both biotic and abiotic factors influence the distribution and diversity of reduviid predators. In agroecosystems, abundance of this predator has been artificially enhanced and makes them sustain; for instance, the influence of mulching and intercropping on the abundance of the Reduviid predator, *Rhynocoris fuscipes* (Claver and Ambrose, 2003).

Many authors studied the pest suppression efficacy of *R. fuscipes* on various crop pests (Singh and Gangrade, 1975; Ponnamma *et al.,* 1919; Singh, 1985; Ambrose and Livingstone, 1986b; Singh and Singh 1987; Ambrose, 1995), particularly *Riptortus clavatus* Thunberg (Heteroptera: Alydidae) (Ambrose, and Claver, 1995), *Spodoptera litura* F. (Ambrose, and Claver, 1995), three pests of pigeonpea (*Cajanus cajan*) (Ambrose, and Claver, 2001), *Helicoverpa armigera* (Hubner), *Nezara viridula* (L.) and *Riptortus clavatus* Thunberg (Claver and Ambrose, 2003) and reproductive performance on three lepidopteran pests (Babu *et al.,* 1995; Ambrose and Claver, 1995; George and Ambrose, 1999b, c; George *et al.,* 2002) has also been studied. Claver and Ambrose (2001) evaluated suitability of various substrata for the mass rearing of this reduviid. Moreover, field release studies showed its efficacy in pest management programme (Claver and Ambrose, 2003a and 2003b).

1.5. ROLE OF REDUVIID PREDATORS IN AUGMENTATION PROGRAMME

Prey record of reduviids is large and diverse; conservation and augmentation of the reduviid predator and its utilization in biological control of the insect pests has been gaining momentum in India (Joseph, 1959; Lakkundhi and Prashad, 1987; Sitaramiaha and Sathyanarayana, 1976; Ambrose, 1995, 2001; Navarajan Paul, 2003; Sahayaraj, 1999, 2007a) and other countries in recent years (Schaefer, 1988; Grundy and Maelzer, 2000; Fadare and Amusa, 2003; Marques et al., 2006; Sachkett et al., 2007; Grundy, 2007). Though conservation and augmentation are two different theoretical phenomena, they cannot be taken apart because, augmentation usually produces effects that are interrelated to each other (Rabb *et al.,* 1976). Conservation and augmentation of the reduviids can be achieved by manipulation of the natural enemies (De Bach and Hagen, 1991) with abiotic factors (Chapman, 2000) in order to make them more efficient in the management of the pest population.

Augmentation, or accelerated production of the biological control agents at roughly one million times the female progeny rate during the time required for the completion of one generation of biocontrol agents with economical procedure involving minimum labour, is a prerequisite for any successful biocontrol programme (Clark *et al.,* 1978). Edward attempted augmentation of the reduviid predator as early as 1962. In *Rhynocoris carmelita* (Stal.) and *Platymeris rhadamathar* (Gerstalker) (Rhyckman and Rhyckman, 1996); *Reduviius sensiles,* (Faust) *Reduviius vanduzueri* (Wygodizinsky and Usinger,) and *Reduviius sonoraensis* (Walker); Grundy (2004); Grundy Maelzer (2002a, 2002b). Furthermore, Tawfik *et al.* (1983a, 1983b) also recorded the augmentation behaviour of *Allaeocranum biannulipes* (Montr and Singh.)

In India, an exotic reduviid predator *Platymeris laevicollis* (Distant) was colonised in a laboratory and released in large numbers at the crowns of the coconut trees at Pandalan in Kerala and Androth in Lakshadweep and Vital in Karnataka (Antony *et al.,* 1979). They found the establishment of the predator population and controlled the *Orius rhinocerous* beetle. For the past two decades Sahayaraj (1999), Sahayaraj and Martin (2003), Sahayaraj and Balasubramanian (2008), Sahayaraj and Ravi (2007), Ambrose (1988, 1995), Claver and Ambrose (2003a, 2003b), Schaefer (1988), and Sahayaraj (2006) feel that there is an urgent need for evolving strategies to mass rear the potential reduviid predators and their subsequent large scale of release into the pest infested agroecosystem and to assess its biological control potential.

1.6. NECESSITY TO STUDY THE TEMPERATURE IMPACT TO REDUVIID PREDATORS

Studies are currently being conducted on various aspects of the biology and ecology of reduviid predators to better understand their role as a natural enemy of many economically important insect pests. An important issue in ecological theory is to understand how increased or reduced temperature levels associated with global climate change will affect ecosystems, and in particular, trophic interactions. Impacts of climate change on organisms are already apparent, with effects ranging from individual to ecosystem scales. Recent studies indicate that previous forecasts were conservative in their predictions for the magnitude of global warming. Up-to-date models suggest that the global mean surface temperature will increase by 1.8 to 4°C by the year 2100. The ecological impact of such warming is already apparent in the effects seen on species fitness, range shifts, species interactions, and community structure. Temperature is a critical abiotic factor influencing the dynamics of mites and insect pests and their natural enemies. Temperature sets the limit of biological activities of arthropods, such that low and high temperature thresholds and optimal temperature can be estimated for all major life processes. The effect of temperature on rate of development and survivorship is of fundamental importance to understanding insect phenology and abundance. Such data are also necessary to gain an understanding of interactions between prey and natural enemies in biological control systems. Microclimatic conditions such as temperature, humidity and soil moisture lead to variation in the susceptibility of conspecific preys to carnivore-like reduviid predators. Under field sitation soil-dwelling reduviids concealed under small builders, whereas arborial reduviids preferred to stay underneath the leaves. However, these predators are having lots of chances to be exposed to sunlight. Sunlight exerts a positive effect on the growth of many insects, usually through the influence of temperature on the rate of consumption of prey. However, a species' physiology and thermoregulatory behaviour determine whether or not its growth will be sensitive to variation in temperature (Knapp and Casey, 1986).

Little has been published on the effect of temperature on development and survivorship in reduviids. Shepard For the first time, James (1992) studied the impact of various constant temperatures (22.5–35°C) on the biology of *Pristhesancus plagipennis* Walker. *Rhynocoris marginatus* and *R. fuscipes* distributed all over India. At the end of South East Monsoon (October to December), the northern part of India is very cold, whereas the southern part is much hotter. Thus, a great deal of variation exists in both pest injury and predatory efficiency across the geographic range of these species that likely reflects variation in their biological performance under local conditions, especially their responses to temperature. Furthermore, when a predator is used as a biological control agent, the suitability of the habitat for multiplication and development is of paramount importance. Little has been published on the effect of temperature on development and survivorship in reduviid predators. No information is available about the impact of temperature on the physiology of reduviids. Hence, to study the impact of temperature on the biology and physiology is worthwhie to discuss in detail in the book. We discussed the impact of temperature in relation to the following aspects:

i.   Life history
ii.   Lower (LDTT) and higher (HDTT) developmental threshold temperatures
iii.   Stage preference against hemipteran and lepidopteran pests
iv.   Biological control potential
v.   Egg and adult macromolecules
vi.   Gut microbial profiles and their hydrolytic enzymes production
vii.   Atigen and antibody interaction by ELISA
viii.   DNA amplification and genetic similarity

Chapter 2

LIFE HISTORY AND THRESHOLD TEMPERATURES

ABSTRACT

The duration of nymphal developmental period, reproduction and longevity of *Rhynocoris marginatus* and *Rhynocoris fuscipes* were studied in the laboratory at six constant temperatures (10-35°C). At 10, 15 and 35°C eggs had not developed completely and sustained 100 and 75% of mortality in each temperature regimes. Similarly, life stages of these two predators didn't reach the adult stages particularly at the temperatures of 10-15 °C and 35°C. Among the different temperatures tested, the total nymphal development period of *R. marginatus* was maximum at 20°C (87.26 days), reduced at 25°C (49.8 days) and further reduction was recorded at 30°C (48.06 days). When freshly molted adults of *R. marginatus* were subjected to different temperature regimes the longest longevity was observed at 20°C (7.66 and 17.26 days for male and female respectively). The egg of *R. marginatus* hatched within 7.58 days with 95.5% hatchability. Irrespective of the temperature levels, the incubation period got prolonged in *R. marginatus* adults maintained at different temperature levels. The longest and the shortest adult longevity has been observed on both male and female predators such as 28.66 and 69.75 days at 20°C, although, in *R. fuscipes* the longest longevity (35.25 and 32.5 days for male and female respectively) was recorded at 25°C. Results from this study were used to enhance the efficiency of mass-rearing methods in the laboratories, as well as field releasing purposes.

2.1. INTRODUCTION

Influence of temperature on life history and bioefficacy of other predatory hemipteran predators like Lygaeidae (Dunborn and Bacon, 1972), Pentatomidae (Mukerjii and Roux, 1965, Goryshin and Tuganova, 1989; Westich and Hough-Goldstein, 2001; Legaspi, 2004), and Anthecoridae (Isenbour and Yeargan, 1981, Izumi and Ohto, 2001; Kohno and Kashio, 1988; Parajulee and Philips, 1992) were available in the literature. *Rhynocoris marginatus* is a polyphagous, multivoltine, entomosuccivores, polymorphic, crepuscular and alate bug predominantly found in the scrubjungles, semi arid zones, tropical rain forests and agroecosystems of south India (Livingstone and Ambrose, 1978; Sahayaraj, 1994 and 2002a, b, 2007a). It is a selective biological control agent of many agricultural and forest insect pests like *Earias fraterna* (Fab.) (Ambrose, 1988), *Papilio demoleus* (L.) *Earias vittella* (Fab.) (Nayar *et al.,* 1976), *Corcyra cephalonica* (Stainton) (Bhatnagar *et al.,* 1983), *Helicoverpa armigera* (Hubner) (Ambrose 1987), *Mylabris indica* (Faust), *Mylabris pustulata* (Fab) (Imms, 1985 and Nayar *et al.,* 1976), *Achaea janata* (Linn.), *Oxycarenus hyalinipennis* (Costa) and *Approarema modicella* (Deventer) (Sahayaraj, 1995a, b; Sahayaraj *et al.,* 2003) and *Spodoptera litura* (Fab.) and *Amsacta albistriga* (Walker.) (Sahayaraj, 2000).

Recently, Sahayaraj (2007a) recorded the biological control potential of *R. marginatus* on four groundnut pests under laboratory condition. George *et al.* (2002) observed the nutritional influence of prey on the biological control and biochemistry of *R. marginatus*. There was a great effect of biopesticides like Neem gold and nuclear polyhydrosis viruses (NPV) on the incubation period and hatchability of *R. marginatus* eggs (Sahayaraj and Paulraj, 1999). Previously, the impact of space (Vennison, 1988), mating behaviour (Ambrose and Livingstone, 1985), starvation (Ambrose *et al.,* 1990a, 1990b) and prey influenced (Ambrose and Claver, 1996) the biology of the bug. Bio-efficacy and prey size influenced the developmental period (Sahayaraj, 1995a, 1995b, 1995c) and predatory potential of the bug (Ambrose and Claver, 1996). Ecotypic (Ambrose, 1987) and polymorphic diversity (Ambrose and Livingstone, 1978; Vennison and Ambrose, 1988) of the reduviid was also documented. Sahayaraj *et al.* (2003) observed the effect of two biopesticides namely neem gold and NPV on the eggs and nymphal instars of the predator. Similarly, Sahayaraj and Martin (2003) witnessed augmented control in groundnut pests. Recently Sahayaraj et al. (2007a, 2007b) and Sahayaraj and Balasubramanian (2008) studied the prey influence on salivary gland and gut enzyme profiles of the reduviid.

Rhynocoris fuscipes is a crepuscular, brightly coloured (black and red), entamophagous, harpactorine reduviid found in concealed habitats such as underneath the stones and in crevices (Ambrose and Mayamuthu, 1994; Ambrose, 1987; George *et al.*, 2000). When it is found in an agroecosystem, it predates upon various insects like *H.armigera* (Hubner), *C. cephaionica* (Stainton), *A.janata* (Linn.), *Plutella xylostella* (L.) *S. litura* (Fab.) *Myzus persica* (Sulz.), *Lygus hespes* (Fabricius), *Rhaphid opaipa* (Thunb.), *Foveicollis lucas* (Distant.), *Semiethisa pervolagata* (Walker.) (Singh, 1985), *Epilacrisia stigma* (Muls.) (David and Natrajan, 1989), *Cryptosilla pyranthes* (Linn.) (Hiremath and Thondarya, 1983), *Calocoris angustatus* (Leth.) (Ambrose, 1980), *Patanga succincta* (Linn.), *Dysdercus cingulatus* (Fab.), *Earias vittella* (Fab.) (Singh and Singh, 1987), *Nezara viridula* (Linn.) (Singh and Gangrade, 1975), *Perigrinus maidiis* (Ashm.), (Ponnamma *et al.*, 1919), *Spilosoma obliqua* (Walker.) (Cherian and Kylasam, 1939), *Myllocoris curvicornis* (Fab.), (Cherian and Brahmachari, 1941); *Aulacophosa foveicollis* (Fab.), (Ambrose, 1995); *Pleopidas mathias* (Fab.), *Clavigarata gibbosa* (Spinda.), *Clavigarata horrens* (Distant.), *Dolycoris indicus* (Stal.) (Das, 1996). Many authors studied the pest suppression efficacy for mass rearing and functional response of *R. fuscipes* on various crop pests (Singh and Gangrade, 1975; Ponnamma *et al.*, 1919; Singh, 1985; Ambrose and Livingstone, 1986b; Singh and Singh 1987; Ambrose, 1995); reproductive performance on three lepidopteran pests (Babu *et al.*, 1995; Ambrose and Claver, 1995; George and Ambrose, 1999b, 1999c; George *et al.*, 2000) has also been studied.

Proper storage of the natural enemies of pests is essential to face the problems related to production, planning and the unpredictability of demand. Cold storage is a useful technique to ensure the availability of beneficial insects for further research or field release without maintaining or continuous rearing. Furthermore tolerance to cold may be considered as a desirable attribute for shipment procedure. The performance of an insect species depends on many factors including temperature, which affects the rate of development, reproduction and aging. Development rate reproduction and mortality at a range of constant temperatures allows the construction of life tables and reveals the effect of temperature on species performance (Oliveira et al., 2010). The development of a forecasting system for the use of reduviid predators in an IPM programme largely depends on the understanding of the relationship between temperatures and development of the species of interest. Several studies have addressed the effect of temperature on the biology of reduviids such as *Allaeocranum quadrisignatum* (Fab.) (Tawfik et al. 1983a, b), *Pristhesancus plagipennis* (Walker) (James 1992), *Amphibolus venator* (Klug) (Pingale, 1954; Nishi and Takahashi, 2002) and egg hatching of *R. marginatus* and *fuscipes* (Sahayaraj and Paulraj 2001a, b). Further more, tolerance to low temperature may be considered as a desirable attribute in shipment procedures (Anderson et al. 2001). The hunter reduviids *Rhynocoris marginatus* and *Rhynocoris fuscipes* are important polyphagous predators widely distributed throughout India (Ambrose 1995; Sahayaraj 1995a, b). These reduviids have been found in the agroecosystems such as groundnut, cotton, sugarcane and soybean and also in semi-arid zones and scrubjungles (Sahayaraj 2002a, 2002b). To our knowledge however, no attempt has been made to determine the development, survival and fecundity of *R. marginatus* and *R. fuscipes* in relation to temperature. Therefore a study was undertaken to study the effects of constant temperature on development and reproduction of *R. marginatus* and *R. fuscipes*. Moreover, we are going to predict the optimum temperature for rearing these reduviids that could be used for further experimental study of these species.

2.2. METHODOLOGY FOLLOWED

Rhynocoris marginatus and *R. fuscipes* were collected in their life stages from Tirunelveli District, Tamil Nadu, India. They were maintained in the laboratory at $29 \pm 1.5°C$, $75 \pm 5\%$ RH and 11L: 13D photoperiod on fourth and fifth instar larvae of *Corcyra cephalonica* larvae. The eggs laid by the predator were maintained in a small plastic container (60ml capacity). Newly hatched first instar nymphs were used for this experiment. Development, survival and reproduction of both *R. marginatus* and *R. fuscipes* were studied using environmental chambers at six constant temperatures *viz.,* 10, 15, 20, 25, 30 and 35°C. The photoperiods for all experiments were 13D: 11L and relative humidity was maintained in $65 \pm 5\%$. Individual egg batches were placed on filter paper in small plastic vials (60ml volume) with a perforated lid and humidity provided by a piece of wet cotton swabs. The experiment started with egg batches (approximately 100 to 153 in *R. marginatus* and 58 to 90 eggs for *R. fuscipes*) containing 34, 45, 58 and 62 eggs per batch. They were subjected to each treatment regime, replicated thrice, and the vials were kept in the above-mentioned temperatures. Control categories were maintained at room temperature ($28 \pm 0.5°C$) (RT). Egg development was monitored twice daily and their hatching percentage was recorded at each temperature separately. Upon emergence, the nymphs were transferred into plastic boxes (12cm width and 4cm length) and maintained up to the third instar. Then they were transferred to larger plastic boxes (15cm width and 7cm length) and maintained up to their death on *C. cephalonica* larvae in *ad libitum*. Weights of the newly hatched nymphs were recorded using Monopan balance. Upon emergence the sex ratio of the adults was determined (female / female + male). Both weight gain and weight loss of the predators was determined by using the following formula adopted by Arnold (1959).

$$\text{Weight gain/loss} = \frac{\text{Final weight - Initial weight in the treatments}}{\text{Final weight - Initial weight in the control}}$$

Two males and two females were maintained separately in large plastic vials (15 cm width and 17 cm length) until their death with *C. cephalonica* in *ad libitum*. The following was recorded:

- Pre-oviposition period = number of days from emergence and first oviposition;
- Oviposition period = number of days from first to last oviposition;
- Post-oviposition period = number of days from last oviposition to mortality;
- Daily oviposition = number of eggs laid per female per day;
- Total fecundity = total number of eggs laid per female; and
- Per cent egg viability = percentage of eggs hatched.

Water supply was checked in the environmental chambers daily. In another study, newly emerged adults (> 6 hrs old) at a 2:1 ratio (male:female) were also subjected in to the environmental chamber daily and all the above-said parameters were recorded.

Reciprocals of the observed nymphal developmental duration (in days) were considered as developmental rates of each stage. Raw data from the developmental rate was regressed against six experimental temperatures by linear regression analysis. Further, to examine *R. fuscipes* and *R. marginatus* instar–specific thermal developmental unit requirements, developmental models were constructed and degree-day calculations were performed after Braman *et al.* (1992). Lower developmental threshold temperatures (LDT) were calculated by extrapolating the linear regression line to the x-axis of a graph with the reciprocal of developmental time on the Y-axis and increasing temperature on the X- axis (Arnold, 1959). Upper developmental threshold (UDT), the temperature above, which the developmental rate decreased, was estimated directly from the data.

Only the development times of *R. fuscipes* and *R. marginatus* that attained adult stage were used to estimate the lower developmental threshold (To) and upper development threshold (Tb). The mean number of degree-days (DD) required for development of each life stage was calculated using the following equation as suggested by Price (1984):

$$DD = D (T-t)$$

Where D was the developmental duration (days) 'T' was the temperature in °C during development and 't' is the lower developmental threshold in °C.

2.3. OBSERVATIONS

2.3.1. Nymphal Developmental time

Both *Rhynocoris marginatus* and *Rhynocoris fuscipes* completed their nymphal development when they were reared between 20 and 30°C. However, the total nymphal developmental period of both reduviids decreased with increasing temperature from 20 to 30°C. At 35°C the nymphs were developed faster and died during the beginning of the third nymphal instar. Among the different temperatures tested, the total nymphal development period of *R. marginatus* was highest at 20°C, less at 25°C and further reduction was recorded at 30°C (Table 1). It was slightly decreased when *Rhynocoris marginatus* was maintained under room temperature. A comparison between 20 and 25°C (P=0.0739), 25 and 30°C (P=0.0713), and 20 and 30°C (P=0.0049), shows that room temperatures with 20°C (P=0.0069) and 30°C (P=0.0608) were significant.

A similar trend was also observed in *Rhynocoris fuscipes*. For instance, at 20°C, the total nymphal developmental period was prolonged 37 days compared with room temperature (Table 1). When these predators were maintained at 10°C and 15°C, they had not reached the adult stage. Fluctuation of the temperature above 35°C and below 10°C and 15°C, resulted in none of the eggs or nymphs hatching or closing successfully in both the predators; during moulting at high temperature (35°C) and also at lower temperatures (10°C and 15°C) nymphal stages were unable to shed their exuviae and the insects eventually died. Although statistical significance was recorded between 20 and 25°C (P=0.0084), control to 20°C (P=0.0077) and between control to 30°C (P=0.0851), the comparison between 20 to 30°C was insignificant (P = 0.0131) at 5% level.

Table 1. Nymphal total developmental time (in days) (NTDT) and survival rate (in %) of the reduviid predators

Temperatures (°C)	Nymphal total developmental time		Nymphal total survival rate	
	R. marginatus	*R. fuscipes*	*R. marginatus*	*R. fuscipes*
10	43.0 (I-II)	42.3 (I-II)	18.2 (I-II)	15.4 (1-II)
15	57.1 (I-III)	54.3 (I-II)	38.8 (I-II)	16.9 (I – IV)
20	87.2	82.0	34.3	32.0
25	49.8	47.4	52.3	49.8
30	48.0	44.6	58.6	50.1
35	9.4 (I-II)	9.6 (I-II)	20.8 (I-IV)	16.35 (I-III)
RT	46.0	45.5	88.3	83.25

2.3.2. Egg and nymphal survival rate

At temperatures between 10 and 20°C, in *Rhynocoris marginatus,* hatching or egg survival ranged from 68.2 to 100% (Table 1). In *Rhynocoris fuscipes* egg survival was highest at 30°C and lowest at 10°C, followed by 25°C (Table 1). At a higher temperature (35°C), the nymphal survival rate was arrested after fourth and third nymphal instars for *Rhynocoris marginatus* and *Rhynocoris fuscipes,* respectively. Between the two predators, the nymphal survival rate of *R. fuscipes* was higher than *Rhynocoris marginatus.* When the nymphs and adults were subjected to six constant temperatures, cannibalism occured both in nymphal stages and adults at 35°C. Maximum and minimum nymphal total survival rate was recorded in room temperature and 20°C respectively for both reduviids (Figure 1).

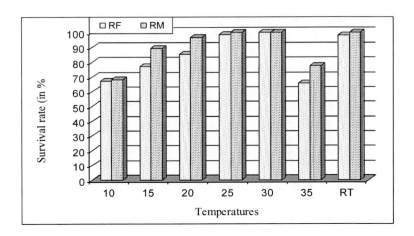

Figure 1. Temperature dependent egg total survival rate (in %) of *R. fuscipes* (RF) and *R. marginatus* (RM)

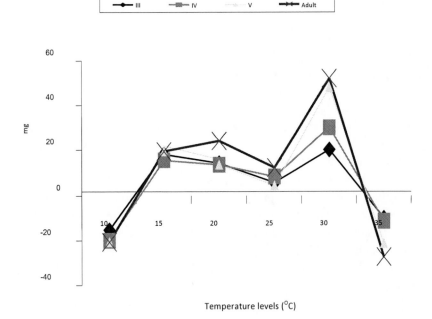

Figure 2. Impact of various temperatures (°C) on the body weight loss (-) and gain (+) (in mg.) of *R. marginatus* (A) and *R. fuscipes* (B) nymphs and adults

2.3.3. Sex ratio and Adult Longevity

Sex ratio of *Rhynocoris marginatus* was female biased (0.86) at room temperature. It diminished gradually from 30°C to 20°C (0.82, 0.78 and 0.69 for 30, 25 and 20°C, respectively), whereas in *Rhynocoris fuscipes* the sex ratio was maximum at 30°C (0.98) followed by room temperature (0.82) and 25°C (0.68).

When freshly moulted adults of *R. marginatus* was subjected to different temperature regimes the longest longevity was observed at 20°C, 7.66 and 17.26 days for male and female, respectively. Similarly in *R. fuscipes*, the longest longevity was observed at 20°C for male (73.33 days) and female (90.73 days) followed by 15°C and 10°C; significantly shorter longevity was observed at 35°C. The longest and the shortest adult longevity have been observed on both male and female predators as 28.66 and 69.75 days at 20°C. In *R. fuscipes,* the longest longevity (35.25 and 32.5 days for male and female, respectively) was recorded at 25°C. Longevity of the predators recorded both at 25 and 30°C indicated similar life span when compared to room temperature for *R. marginatus*. Whenever temperature increased, longevity was also affected or decreased in both predators. Of these two tested predators, *R. fuscipes* had a high, tolerant capacity at lower temperatures of 15 and 20°C.

2.3.4. Reproduction

Tables 2 to 5 show the oviposition pattern of *R. marginatus* and *R. fuscipes* in relation to different temperatures. In *R. marginatus* the pre-oviposition period was 12.9 days at room temperature, but it was prolonged to nearly two months (58.1± 1.9days) when freshly moulted adults were subjected at 20°C (Table 2). In general, the preoviposition period diminished when the temperature increased. Oviposition period lasted for two months (57.2±5.0 days) at room temperature. It was gradually but significantly (p=0.0026) diminished when the temperature was increased from 20 to 35°C (Table 3). Statistical comparison between the pre-oviposition and post-oviposition periods (20-25°C) was highly significant (F= -7.91229; df= 15; P=0.0010); also, within the oviposition periods between 25-30°C was significant (P< 0.05). However the comparison of oviposition period between 25°C to control category was insignificant.

In *Rhynocoris fuscipes*, the pre-oviposition period was the same both at 30°C and at room temperature (21.33±4.54 days). However, it prolonged to more than 2 weeks when *R. fuscipes* was maintained at 25°C. Oviposition period was shorter at 30°C than at RT and 25°C in *R. fuscipes*. Post-oviposition period increased gradually from room temperature to 30 and 25° C (Table 4). Oviposition period lasted for 57.1 days at room temperature; it was highly reduced (2.09±0.43 days) at 35°C, followed by 30, 25 and 20°C in *R. marginatus*. Statistical analyses for both pre- and post-oviposition periods revealed 20-30°C and 25-30°C were significant (F= 1.07; df= 10; P<0.05). In *R. fuscipes* the oviposition period was shorter than the pre-oviposition period except at 25°C and 30°C. Statistical comparison between pre-oviposition and post-oviposition between 25-30°C (F= 1.40128; df= 12; P=0.0089) and 20-25°C (P=0.0014) was found to be significant. Within the post-oviposition the temperatures between 30 to 25°C (P= 0.0987) and 30°C and control (P=0.0102) were also significant.

Table 2. Influence of different temperature levels (°C) on reproductive parameters of freshly moulted *R. marginatus* adults

Ovipositional parameters	Temperature levels (°C)				
	20	25	30	35	RT
Preoviposition period (days)	58.1 ± 1.9*	19.6 ± 6.8*	4.80 ± 4.2*	4.6 ± 1.8 [ls]	12.9 ± 0.9
Oviposition period (days)	27.1 ± 5.4*	26.2 ± 9.4 [ls]	21.53 ± 9.2*	12.6 ± 5.0*	57.2 ±_ 5.0
Post oviposition period (days)	13.7 ± 4.01 [ls]	8.2 ± 3.6 [ls]	8.6 ± 0.6*	2.11 2.3 [ls]	18.1 ± 0.7
Total no. of eggs	112.0 ± 6.3 [ls]	154.3 ± 8.6*	175.0 ± 5.5*	147.0± 5.5*	119.6 ± 8.5
Maximum no of eggs/batch	66.7 ± 5.3**	80.3 ± 11.2*	98.0 ±13.8 [ls]	89.0± 13.7 [ls]	78.5 ± 9.1
Minimum no of eggs/batch	31.7 ± 3.0 [ls]	74.0 ± 18.3*	77.0 ± 5.2*	29.0 ± 5.2**	36.7 ± 4.3
Incubation Period (days)	19.2 ± 1.8**	8.1 ± 1.4*	8.49 ± 1.3*	5.44 ± 1.1*	7.8± 0.5
Hatching (%)	9.6*	97.5*	94.7*	74.7	95.5
Egg Mortality%	10.46*	2.6*	5.3 [ls]	26.32**	1.09
Oviposition Index	3.79	3.19	2.50	5.98	3.16

** = P<0.005 significant
*= P<0.01 significant at 1% level
Is = insignificant at 5% level

Rhynocoris marginatus laid a mean of 119.6 eggs per female. It was enhanced when the predator was subjected to 30°C. The egg production was high at 25°C. However, when a freshly emerged *R. marginatus* was subjected to different temperature regimes, the egg production decreased from the control category to 25 and 30 °C. When *R. marginatus* was maintained at different temperature levels from the first instar to adult, the egg production decreased gradually from 30 to 20°C. A similar trend was also recorded in *R. fuscipes*. Statistical comparison between control and 20°C (P = 0.0418), 20-25°C (P = 0.0737), 20- 30°C (P = 0.049), and 25-30°C (P = 0.0713) were significant in *R.marginatus*. Comparison between control to 20°C (P=0.0007), 25°C (P=0.0611), 30°C (P=0.0851) and also between 20 to 25°C (P=0.0084) were significant at 5% level of paired sample't' test. Both minimum and maximum number of egg per batch per female was also recorded (see table 4 and 5). *Rhynocoris marginatus* laid a maximum number of eggs at 30°C (175.0 eggs /batch /female) when adults were subjected to this temperature. It was significantly reduced at 25°C (154.31 egg/female) followed by 35°C (147.0 egg/female) and 20°C (112.0 egg/female).

Table 3. Reproductive parameters of freshly moulted adult *R. fuscipes* due to various temperature levels (°C)

Ovipositional parameters	Temperature levels (°C)			
	20	25	30	RT
Preoviposition period (days)	0	36.02 ± 7.41	21.33 ± 5.10 ^{ls}	21.3 ± 4.5
Oviposition period (days)	0	22.78 ± 16.21*	15.06 ± 2.79*	20.6 ± 3.3
Post oviposition period (days)	0	11.34 ± 4.20 ^{ls}	5.83 ± 2.26*	4.8 ± 2.2
Total no. of eggs	0	98.36 ± 7.41*	73.72 ± 3.55**	229.7 ± 5.1
Maximum no of eggs/batch	0	61.50 ± 10.80*	65.02 ± 3.09 ^{ls}	221.1 ± 6.7
Minimum no of eggs/batch	0	29.07 ± 2.13 ^{ls}	35.07 ± 1.08 ^{ls}	8.03 ± 1.2
Incubation Period (days)	22.17 ± 0.53**	9.4 ± 1.47*	8.20 ± 1.0*	8.66 ± 0.7
Hatching (%)	71.42**	88.23 ^{ls}	96.71*	92.8
Egg Mortality%	28.38**	11.77 ^{ls}	3.29*	7.2
Oviposition Index	0	2.80	2.11	2.83

** = P<0.005 significant

*= P<0.01 significant at 1% level

Is = insignificant at 5% level

The egg of *R. marginatus* hatched within 7.58 days with 95.5% hatchability. Irrespective of the temperature levels the incubation period got prolonged in *R. marginatus* adults maintained at different temperature levels. However, the percentage of hatchability reduced at 20°C, was statistically insignificant at 25°C and significantly (df=-3.47804; F= 13; P = 0.0220) enhanced at 30°C. The egg of *R. fuscipes* hatched in 7.44 days with nearly 100% hatching in control category. It became insignificantly (P = 0.1030) reduced both at 25°C and 30°C. In contrast to the above observation, *R. fuscipes* egg production was high when the adults were subjected to 25 and 30°C. Similar observation was also recorded for both maximum and minimum number of eggs per batch, incubation periods and hatchability. When compared with other temperature regimes, incubation periods enhanced significantly at 20°C. Egg hatching percentage increased from 35 to 30°C and 25°C. Minimum egg hatchability was recorded at 20°C. A similar trend was also observed in *Rhynocoris fuscipes* when the adults were subjected to different temperature regimes.

Table 4. Reproductive parameters of *R. marginatus* due to various temperature levels (°C)

Reproductive Parameters	Temperature levels (°C)			
	RT	20	25	30
Preoviposition period (days)	12.9 ± 0.9	30.28 ± 12.4*	19.25 ± 2.5*	19.75±6.49
Oviposition period (days)	57.2 ± 4.0	16.50 ± 4.7[Is]	17.00 ± 9.8*	12.83 ± 3.7
Post Oviposition period (days)	18.1 ± 0.7	23.50 ± 7.5*	26.75 ± 3.2[Is]	29.5 ± 5.4
Oviposition index	0.50	0.51[Is]	0.65**	0.44**
Incubation period (days)	7.58 ± 1.5	16.03 ± 2.6*	9.25 ± 12.9*	8.5 ± 0.9
Maximum no of eggs/batch	128.3 ± 17.2	72.08 ± 5.26[Is]	114.03 ± 10.5**	120.07 ± 9.6*
Minimum no of eggs/batch	69.80 ± 14.31	18.39±11.5**	32.60 ± 17.3*	58.39 ± 9.09**
Total no. of eggs	201.30	91.25 ± 1.6[Is]	146.08 ± 3.1*	180.36 ± 5.11
Hatching (in%)	95.5	89.40	96.09	98.23

RT- Room temperature

** = P<0.005 significant

*= P<0.01 significant at 5% level

Is = insignificant at 5% level

Table 5. Reproductive parameters of *R. fuscipes* maintained on various temperature levels (°C)

Reproductive parameters	Temperature levels °C		
	25	30	Control
Preoviposition period (days)	6.80 ± 3.4	7.60 ± 1.51	4.83 ± 2.26
Oviposition period (days)	11.75 ± 4.4 [Is]	13.05 ± 2.2	20.66 ± 3.55
Postoviposition period (days)	$18.20 \pm 5.3*$	6.50 ± 1.9	21.48 ± 4.52
Maximum no. of Eggs / batch	$68.26 \pm 17.7*$	$75.8 \pm 14.5*$	79.71 ± 5.27
Minimum no. eggs / batch	29.16 [Is]	$43.52**$	47.16
Total no. of egg / batch	$92.52 \pm 2.84*$	$118.11 + 2.73**$	120.89 ± 1.31
Oviposition index	$0.65*$	0.36	0.90
Incubation period (days)	9.20 [Is]	$8.11*$	7.44
Hatching (in %)	$98.20*$	$98.81*$	99.61

** = P<0.005 significant; *= P<0.01 significant at 1% level; Is = insignificant at 5% level

2.3.5. Developmental Models – Linear Model

Table 6. Linear thermal unit models for lower threshold temperatures (T_o) (°C) and mean thermal unit recruitment (K) on the development of the nymphal instars of *R. marginatus* and *R. fuscipes*

Life Stages	R. marginatus				R. fuscipes			
	Regression equation	r^2	To	K : DD	Regression equation	r^2	To	K : DD
Egg	$Y = 0.0267\,t - 0.2591$	0.018	19.51	52.4	$Y = 0.0145\,t - 0.1462$	0.2820	20.53	67.55
1st instar Nymph	$Y = 0.0422\,t - 0.5610$	0.021	21.16	65.6	$Y = 0.0327\,t - 0.3230$	0.1723	19.41	64.71
2nd instar Nymph	$Y = 0.0394\,t - 0.5270$	0.481	16.92	73.6	$Y = 0.0512\,t - 0.6410$	0.4531	22.48	72.60
3rd instar Nymph	$Y = 0.0433\,t - 0.3114$	0.258	12.09	60.4	$Y = 0.0659\,t - 0.6337$	0.8472	15.58	63.1
4th instar Nymph	$Y = 0.860\,t - 0.3526$	0.525	18.05	43.51	$Y = 0.0996\,t - 0.3253$	0.528	14.63	54.0
5th instar Nymph	$Y = 0.0229\,t - 0.4611$	0.810	18.40	41.4	$Y = 0.02425\,t - 0.1253$	0.635	12.07	52.96
I – Adult	$Y = 0.0058\,t - 0.0691$	0.851	24.31	284.5	$Y = 0.0511\,t - 0.0528$	0.570	25.01	289.67
Total	$Y = 0.0031\,t - 0.0590$	0.140	24.8	336.0	$Y = 0.0062\,t - 0.0674$	0.756	23.92	362.76

Y = reciprocal of mean developmental times; r^2 = coefficient of correlation; Degree-days (DD) requirement to complete instar.

Both linear and non-linear models were used to describe the relationship between development rates (L/d), and temperatures. To select the most appropriate model for both predators' development, prediction, a comparative analysis model was performed here. The overall coefficient of *Rhynocoris marginatus* and *Rhynocoris fuscipes* was 0.949 and 0.967, respectively. The linear model adequately described the lower temperature threshold (t_b) and degree-day required for development (Table 6). The linear regression was applied to know the relationship between development rates and temperatures. It was linear for all stages except in the first and second nymphal instars of R. *marginatus* and R *fuscipes,* respectively. The estimated lower and higher temperature thresholds (t_b) for eggs were 19.51 and 20.53 respectively. This threshold level was enhanced for the nymphal stages at lower (24.1°C) and higher temperatures (32.1 °C) for R. *marginatus.* Similarly, R. *fuscipes* revealed the threshold level to be 25.41 and 34.28 for low and high temperature levels, respectively. When we consider the individual nymphal instars, lower and higher thresholds (Table 7) were recorded for fifth and first nymphal instars of *Rhynocoris fuscipes.*

Table 7. Linear thermal unit models for higher threshold temperatures (T_o) (°C) and mean thermal unit recruitment (K) on the development of the nymphal instars of *R. marginatus* and *R. fuscipes*

Life Stages	R. marginatus				R. fuscipes			
	Regression equation	r^2	To	K : DD	Regression equation	r^2	To	K : DD
Egg	Y = 0.0267 t– 0.2591	0.830	24.5	52.45	Y = 0.0145 t – 0. 1462	0.678	30.11	67.50
1st instar Nymph	Y = 0.0622 t– 0. 5610	0.671	23.16	65.6	Y = 0.0327 t – 0. 3280	0.539	20.10	64.71
2nd instar Nymph	Y = 0.0394 t – 0.527	0.655	35.92	73.6	Y = 0.0512 t – 0.6410	0.611	21.49	71.60
3rd instar Nymph	Y = 0.0433 t – 0.3114	0.497	33.09	73.6	Y = 0.0659 t – 0.6337	0.742	25.7	63
4th instar Nymph	Y = 0.0433 t – 0. 3526	0.572	31.05	60.4	Y = 0.0996 t – 0.6517	0.832	31.63	44.0
5th instar Nymph	Y = 0.0229 t – 0.462	0.568	35.27	41.4	Y = 0.071 t – 0. 8315	0.851	30.38	44.26
I – Adult	Y = 0.0053 t – 0.690	0.842	32.31	284.51	Y = 0.0511 t – 0. 0528	0.721	34.28	284.67
Total	0.0031 t – 0.0590	0.788	24.8	358.07	0.0062 x – 0. 0678	0.764	25.3	316.03

Y = reciprocal of mean developmental times; r^2 = coefficient of correlation; Degree-days (DD) requirement to complete instar.

2.3.6. Growth and morphogenesis

Growth of the insect has been recorded in terms of weight gain or loss and size of the predators. Impact of temperature on the total body weight gain and loss in *R. marginatus* and *R. fuscipes* is depicted in Tables 7 and 8. Both at higher (35°C) and lower temperatures (10, 15°C), a remarkable range of weight loss was noticed. At optimum temperature, like 25°C and 30°C, appreciable weight gain was observed with both the predators. Maximum weight loss was recorded at 35°C throughout the experimental periods.

Quantitative measurements of various body parts such as head, thorax and abdomen and the appendages of the two predators at different constant temperature levels are presented in Tables 7 and 8. Low temperature decreased the size of legs, thorax, wings and rostrum of *R. marginatus* (Table 7). But temperature had positive influence on antenna size. However, in *R. fuscipes* size of all the body parts was invariably high at 20°C (Table 8). Deformities were observed both in the nymphal instar and adult of *R. marginatus* and *R. fuscipes*. Predators were reared at 15°C, moulted incompletely and immediately after a few hours the adults had died. Fifth nymphal instars had visibly longer hind legs and tibiae were shorter than the normal nymphs. At 35°C, fore and hind wings of the developed predators were reduced with a fully burnt appearance. Both pairs of wings were incompletely developed and irregular in shape (20°C).

Adults of *R. fuscipes* had permanently attached moulted skin towards the posterior parts until death of the predator. At 20°C, fore and hind wings were very thin change the original position; also whole wings were replaced from that surface. When the nymphs three turned to whole body colour they became pale brick to dark red; such colour was noticed on both species of predator. Since old exuvia was not removed from the body properly, most of the fifth nymphal instars of *R. marginatus* were unable to moult into adults. In *Rhynocoris fuscipes,* the exuvia that remained was attached on the posterior abdominal region. However in *Rhynocoris marginatus* the moulted skin became attached to the legs.

Table 8. Impact of constant temperatures (°C) on the morphometry of head, thorax, abdomen (in mm.) and parts of R. fuscipes adults

Temperature levels (°C)	Antennae	Rostrum	Head	Thorax	Abdomen	Fore Leg	Mid leg	Hind leg	Fore wing	Hind wing (in mm)	Abdominal width (in mm)dth	Thoracic width (in mm)	Total Length (in cm)
20	0.46 + 0.073	0.27 + 0.011	0.29 + 0.02	0.36 + 0.05	0.81 + 0.06	0.76 + 0.04	0.60 + 0.05	0.86 + 0.05	0.88 + 0.08	0.68 + 0.05	0.47 + 0.25	0.23 + 0.02	1.46
25	0.46 + 0.06	0.22 + 0.01	0.25 + 0.48	0.35 + 0.78	0.67 + 0.05	0.68 + 0.053	0.57 + 0.74	0.71 + 0.12	0.74 + 0.07	0.62 + 0.05	0.42 + 0.03	0.23 + 0.02	1.27
30	0.25 + 0.10	0.25 + 1.08	0.24 + 0.73	0.35 + 0.82	0.61 + 0.12	0.68 + 0.15	0.60 + 0.15	0.70 + 0.26	0.54 + 0.12	0.46 + 0.07	0.42 + 0.08	0.20 + 0.02	1.23
R T	0.25 + 0.10	0.24 + 0.17	0.25 + 0.21	0.35 + 0.74	0.65 + 0.23	0.69 + 0.03	0.59 + 0.20	0.70 + 0.09	0.54 + 0.06	0.45 + 0.05	0.43 + 0.03	0.21 + 0.07	1.25

Table 9. Impact of constant temperatures (°C) on the morphometry of head, thorax, abdomen (in mm.) and parts of *R. marginatus* adults

Temperature levels (°C)	Antennae (in mm)	Rostrum (in mm)	Head (in mm)	Thorax (in mm)	Abdomen (in mm)	Fore Leg (in mm)	Mid leg (in mm)	Hind leg (in mm)	Fore wing (in mm)	Hind wing (in mm)	Abdominal width (in mm)	Thoracic width (in mm)	Total length (in cm)
20	0.60 ± 0.09	0.29 ± 0.23	0.26 ± 0.04	0.47 ± 0.04	0.58 ± 0.039	0.79 ± 0.04	0.57 ± 0.08	0.83 ± 0.08	0.61 ± 0.01	0.54 ± 0.087	0.50 ± 0.11	0.34 ± 0.05	1.30
25	0.66 ± 0.17	0.37 ± 0.06	0.31 ± 0.60	0.83 ± 0.12	0.38 ± 0.06	1.02 ± 0.08	0.75 ± 0.10	1.11 ± 0.19	0.94 ± 0.15	0.71 ± 0.12	0.54 ± 0.10	0.35 ± 0.05	1.52
30	0.56 ± 0.20	0.71 ± 0.09	0.35 ± 0.05	0.832 ± 0.18	0.375 ± 0.068	1.19 ± 0.14	0.79 ± 0.07	1.13 ± 0.17	0.97 ± 0.02	0.73 ± 0.149	0.54 ± 0.09	0.34 ± 0.07	1.56
R T	0.56 ± 0.21	0.71 ± 0.08	0.34 ± 0.19	0.84 ± 0.19	0.86 ± 0.062	1.20 ± 0.22	0.79 ± 0.08	1.15 ± 0.19	0.95 ± 0.053	0.77 ± 0.13	0.55 ± 0.07	0.35 ± 0.01	2.08

2.4. DISCUSSION

The ability of a natural enemy to adapt to different environmental conditions is an essential prerequisite for its successful utilization in a biological control programme. Among other environmental conditions, temperature is considered to be a key factor affecting the biology and ecology of both harmful and beneficial insects. The two Reduviid species are distributed in rice, cotton, bendhi, chilli, groundnut, soybean, pumpkin, sugarcane, pigeon pea (Sahayaraj, 2006) and also the adjacent ecosystems like semi-arid zones, scrub jungles and forests throughout India. This indicates that there are a lot of possibilities for exposure to a wide range of temperature between 20-35°C. The temperature levels between 10 and 15°C were chosen in order to find out the possibilities for storing these reduviids for an augmentative release programme. Hence constant temperature treatments between 10-35°C, with 5°C step intervals.

Reduviids are well known generalist predators feeding on a wide variety of prey (Miller 1971). Their value as a regulator of insect pest populations has rarely been investigated. However, recent interest in the extent to which generalist predators that lack density dependent tracking may limit the prey (Murdoch et al., 1985). Very little research work is available about the effect of temperature on nymphal development and survival, and the reproduction of reduviids (James, 1992; Sahayaraj, 2007a).

Development of reduviid predator species viz., R. marginatus and R. fuscipes clearly shows that it increases as the temperature decreases. In contrast, the following authors reported another view that for generalist predators, developmental period of Orius laevigatus (Rudolf et al., 1993) and Lytocoris campestris (Parajulee et al. 1995) increased in accordance with temperature levels. Developmental threshold of both the predator species nymphal instars and other performance shared progressively improved above 15°C (15-20, 20-25 and 25-30°C). Present results confirm the previous reports on other hemipteran predators (Nakamura 2003; Sayaka et al., 2007; Sarah et al. 2007). Nymphal development was generally studied at constant temperatures between 25 and 30°C and about 70% humidity on unspecified conditions of ambient temperature and humidity (Nishi and Takahasi 2002). However, temperature and humidity in insect habitats may differ considerably and vary according to circadian and seasonal patterns (Shinmizu and Kawaski 2001; Logen et al. 2006, 2007). Several authors have reported that acceleration of temperature levels had caused retardation of developmental periods both at fluctuated temperature levels and also at constant temperature levels (Mark and Jervis 2005). Overall, early instar of reduviids required less time to develop than later instars and are consistent with data reported by Neal and Douglass (1988) and Roy et al. (2002) in Tingid, *Stephanitis pyrioides,* Braman et al. (1992) in *Corythuca cydoniae* Usha Rani (1992) in *Eocanthecona furcellatea,* which has been attributed to the higher metabolic activities noticed in the predator (Christian et al. 1999). The results revealed that the survival rate was decreased with increased temperature, which was attributed to absence of brisk activities in the freshly moulted nymphal stages. Similarly, temperature-dependent immature survival and development was also observed from the coleopteran predator (Beckett et al. 1998; Neven 2000). In both reduviids, highest mortality occurred at the time of moulting. Stage specific mortality decreased for successively later stages, demonstrating that early instars of *R. fuscipes* and *R. marginatus* were more vulnerable to thermal stress than later life stages. Therefore, slow development of these reduviids in a cool environment may be due to differences in moulting; that is, during moulting time, incomplete moulting, or failure to shed the outer covering (exuviae), finally leads to death. So survival rate decreased with each instar in accordance with temperature.

Generally, low temperatures slow development and increase the length of periods that the insects are in their various life stages. This can intensify stage-based predation. High temperatures speed development and can cause earlier emergence from the egg stage and a more rapid progression through the nymphal or other stages, giving adults increased periods of reproduction and escape from predation that often accompanies nymphal stages. The predator can experience similar effects. Further, temperature variations can effect daily activity periods, causing either the predator or prey to be more or less active. Therefore, predator search times can be strongly affected by temperature levels. In the present model, the percentage mortality curve versus the whole range of temperatures present an inverted '∩' pattern observed in all the immature stages of *R. fuscipes* and *R. marginatus.* Previously, a similar kind of result was reported with a U shaped curve of mortality versus temperature for the immature stages of coleopterans (Herrera et al. 2005). Survival of both *R. marginatus* and *R. fuscipes* had been remarkably negative at higher temperature. When the reduviids were maintained at 35°C, none of the nymphs reached adulthood. Similarly low temperature levels such as 10 and 15°C were not suitable for the development of these reduviids. These findings are consistent with the previous reports observed in other polyphagous predators (Braman and Pandley 1993). But Diodonet et al. (1996) reported that the temperature shown at the development of predators had a threshold of 10 to 35°C.

Nymphal instars of both *R. marginatus* and *R. fuscipes* obviously have initiated their development, but failed to proceed after the second instar at 10 and 35°C and the third instar at 15°C (only *R. marginatus*), as observed in other heteropteran predators in general and the pentatomid bug in particular by Usharani (1992). However some thriving individuals were unable to leave the old exuvia and died during moulting. This was most prominent when the reduviid was reared at 10 to 15°C. *R. marginatus* and *R. fuscipes* were not able to moult first to second instars at 10°C and second to third and fourth instars at 15°C. A similar kind of observation was also recorded in other reduviid predators (Christian et al. 1999). Both *R. marginatus'* and *R. fuscipes'* nymphal predators were sometimes able to survive for a very long period without undergoing moulting at 15°C (Luz et al. 1998). Denlinger (1991) summarized that the survival was higher for all species of predator at temperatures between 25 - 30°C. Moreover, the nymphal development and growth parameters were significantly affected by day temperatures (Becket and Marton 2003). It may be a reason for the predator not successfully invading in different abiotic stresses.

Sex ratios of *R. fuscipes and R. marginatus* at higher temperatures were found to be female biased. Sex ratio of *R. marginatus* was reported to be female biased under normal laboratory conditions (Sahayaraj et al. 2004). Field collection also shows that both the reduviids were female biased under field situation too. Unfavorable environmental conditions (extreme temperatures, food shortage) may alter the sex ratio of insects (Sarah et al. 2007; Omkar et al., 2009). The observed results indicated that even if the reduviids are present in the unfavorable temperature, it will not alter the sex ratio.

Meanwhile recently, Sahayaraj (2007) evidentially reported that there was a possibility for storing reduviid eggs in the refrigerator for a long period and subsequently using them as a biological control agent as a great benefit to the biological control programme. In this study, introduction of *R. fuscipes* and *R. marignatus* at 20°C prolonged the incubation period for 10 and 12 days respectively. Results show that it can be feasible, if these predator eggs are used in IPM when mass production is required. Hence, it was concluded that *R. marginatus* and *R. fuscipes* could be stored up to 2 weeks with minimum reduction in hatchability at 20°C.

Rhynocoris marginatus and *R. fuscipes* eggs were collected from the laboratory between December and mid January. These eggs failed to hatch, but all the eggs changed to a brownish colour; a similar result was also observed by Hagerty et al. (2001). However, most of the eggs failed to hatch when exposed to either lower or higher temperatures with humidity for shorter and longer periods.

Temperature affects many key life events in phytophagous and zoophagous insects, including their reproductive performance, and particularly fecundity. Carroll and Quiring (1993) found that potential fecundity (number of eggs matured) and realised fecundity (number of eggs laid) are both influenced by temperature. The observed increased in the preoviposition period with a decrease in temperature from 15 to 25C may be ascribed to increased metabolic activity at higher temperature, resulting in more rapid maturation of the gonads. A linear relationship between temperature and rate of gonadal maturation has been described in many phytophagous insects. The extension of embryonic developmental duration at lower temperature results from the depression of egg metabolism due to water loss with low humidity noticed in the eggs of some insects (Frazer and Gregor 1992). This study showed that when the eggs were put on storage at 35°C, most of the eggs failed to hatch. However, eyespots appeared and colour changed with shrunken appearance. This indicates that the similar embryonic development had occurred initially and it did not proceed further as observed by Clark (1996).

Amphibolus venator (Klug) (Reduviidae) is a predator of many stored product insects. It preys on Trogoderma granarium (Everts), Tribolium castaneum (Herbst), Corcyra cephalonica (Stainton), Latheticus oryzae (Waterhouse) and Alphitobius diaperinus (Panzer) (Haines, 1991). A. venator has frequently been found in shipments of groundnuts from Africa to England (Hill, 1990) and also in warehouses in Thailand. Nishi and Takahashi1 (2002) studied the egg production and development of Amphibolus venator (Klug) at five temperatures (25, 27.5, 30, 32.5 and 35°C), 70% r.h. and 16L–8D photoregime. The optimum temperature for multiplication and development of A. venator was around 32.5°C. Pre-oviposition period decreased with increasing temperature. Temperature in the range from 25. °C to 35°C had no effect on oviposition period. The average number of eggs laid per ovipositing female increased with increasing temperatures, however differences were not significant. The percentages of ovipositing females at 32.5 and 35°C were the same (95.8%), but higher than in other temperatures. Total oviposition number was highest at 35°C. Egg incubation period decreased with increasing temperature; however, hatchability was not affected. The development period of nymphs decreased with increasing temperature. The lower developmental thresholds and total effective temperatures were calculated as 16.9°C and 126.6 degree-days for eggs, respectively. The corresponding values for nymphs were 20.7°C and 714.3 degree-days

In the present investigation, results of both the reduviid species showed that low egg hatching was recorded at 20°C. In addition Usharani (1992) defined as a direct response to a limiting factor such as temperature and moisture in which development immediately resumes upon restoration of that factor. In the case of R. marginatus and R. fuscipes results showed that temperature is the limiting factor preventing embryo development and eclosion following oviposition. The present results also showed that the heavier adult predators reared at 25-30°C laid more eggs. A similar phenomenon was reported for a number of other insect species (Frazer and Gregor 1992). It also reveals that reproductive activities of these two reduviid predators were reduced as the temperature increased, which corroborates with the previous reports of Usharani (1992). A similar trend was also observed in the generalist heteropteran predators too (Nagai and Yano 1999; Kino et al. 1999; Ito and Nakata 2000). Lower temperature had a negative effective on the incubation period of the eggs and nymphal survival of the predators studied. Similarly, reduction in the incubation period at high temperature was observed earlier in another reduviid predator *A. biannulipes* (Tawfik and Awadallah 1983 and Tawfik et al. 1983a, b; Torres et al. 1998), neuropteran (Bakthavatsalam et al. 1995) and coleopteran predators too (Omkar and Pervez 2002; Herrera et al. 2005). The incubation of reduviid eggs required little quantity of moisture, prolonged dryness and optimal abiotic factors for the egg development (Vennison and Ambrose, 1990). Furthermore, it was suggested that *R. marginatus* eggs were better able to tolerate too cold temperatures compared to *R. fuscipes* (Sahayaraj and Paulraj, 2003).

Chapter 3

STAGE PREFERENCE AND BIOEFFICACY

SUMMARY

Six constant thermal regimes were compared with room temperature, in order to determine their influence on weight gain (in mg) and predatory rate of the two reduviids. Stage preference studies of *Rhynocoris marginatus* to the life stages of *Spodoptera litura* and *Dysdercus cingulatus* showed that both fifth nymphal instar and adult predators were more successful in encountering large-sized preys. Though different nymphal instars of *R. marginatus* preferred life stages of lepidopteran larvae, second and third instar reduviids preferred second, third and fourth instar *D. cingulatus* nymphs and the remaining life stages of this reduviid often preferred adult *D. cingulatus*. All the nymphal instars and adults of *R. fuscipes* mainly preferred second to fourth instar larvae of *S.litura* and second to fifth instar *D. cingulatus* nymphs. Generally the larger size predator preferred larger prey and smaller size predator preferred smaller preys. Temperatures invariably determine the foraging behaviour of reduviid predators. The variation in biological control with temperature could be described by extended model indicating that temperature influences the attack rate, handling time, predatory rate, and weight gain in adults and nymphs. Biological control potential of *R. marginatus* and *R. fuscipes* on *S.litura* and *D. cingulatus* showed that predators approached their prey quickly at higher temperatures and handled more time. The results provide useful information to establish effective biological strategies with *Rhynocoris marginatus* and *Rhynocoris fuscipes* against *S.litura* and *D. cingulatus*.

3.1. INTRODUCTION

Foods differ in their suitability for an organism, and this simple fact has important implications that ultimately mediate food web interactions, community structure, and ecosystem processes. The relative quality of different foods for select life stages of an animal has been determined for a wide range of trophic guilds represented by diverse animal species, including insects. Both discrete and continuous predator-prey models have formed a fundamental part of ecological theory. However, very few models have included explicit, mechanistic effects of temperature on the interactions. Yet, temperature levels can strongly affect the phenology of both predator and prey, as well as their activity times. Shifts in the phenologies, emergence times, or activity periods can change the timing of the interactions and delay or magnify predation events. These kinds of effects are particularly critical for interaction of insect pests and their predators. The insect pest problem has apparently been aggravated in the world by indiscriminate pesticide application, including aerial spraying which has adversely affected natural bio-researches. Because of the high cost of protecting crops from insect pests with chemical insecticides, the increasing concern over residues in food and gradual depletion of natural resources, there is growing interest in the IPM where natural enemies are an important constituent in the agricultural field. Four species of *Rhynocoris* were reported as potential biological control agents of *D. cingulatus* (Ambrose, 1999; Lakkundi, 1989; Sahayaraj, 2006).

A cotton ecosystem has a variety of natural enemies like pentatomids, reduviids, anthocorids, spiders, ants and coleopteran predators. *Spodoptera litura* is another serious pest of cotton. More than 20 reduviid species were known to predate upon *S. litura* both in the field and laboratory conditions (Sahayaraj, 2007a). The development of forecasting systems for the use of predators in IPM programs largely depends on the understanding of the relationship between temperature and biological control potential of the species of interest. Several studies have addressed the effect of temperature on the development of reduviids (Okasha, 1968a, 1968b). Characterizing temperature responses of predatory insects is an important part of understanding their life history, and is often necessary for predicting their potential as biocontrol agents of herbivorous insects. To our knowledge, however, no attempt has been made to determine biological control thresholds in relation to different constant temperature regimes. Among them *Rhynocoris fuscipes* and *Rhynocoris marginatus* are the common predators, feeding primarily on larval forms of *Spodoptera litura's* life stages and stages of *Dysdercus cingulatus*. Pest suppression efficacy of *R. fuscipes* on various crop pests (Ponnamma et al., 1919; Ambrose 1995; Babu et al., 1995) has been studied. The efficiency of natural enemies is affected by environmental conditions, mainly temperature. Observations presented in chapter 2 show that these predators can develop at temperatures ranging from 20°C to 30°C and that temperature affects the survival of the latter development stages. These biological parameters can vary depending on the species of prey offered. These findings help in the formulation of pest management measures. Generally, there are two main approaches for studying food consumption, utilization, and the relative suitabilities of different foods for animals. First, direct or gravitational methods that weigh the amount of food ingested, weight gained by the organism, and amount of feces produced are useful for estimating consumption and the efficiency at which a species converts ingested material into biomass. Another less labor-intensive approach for assessing a food's suitability is the use of digestive rates of markers (such as chromic oxide or radioisotopes) fed alongside a meal. But there is no study in which *Spodoptera litura* and *Dysdercus cingulatus,* the impotant pests of cotton, are the prey. Therefore our study was undertaken to compare the effect of the temperatures on the predatory rate of *R. fuscipes'* and *R. marginatus'* life stages against *D. cingulatus* and *S. litura*. We considered predatory rate (number of prey consumped/predator/day) and food consumption (weight gained) as tools in this study.

3. 2. METHODOLOGY FOLLOWED

Life stages of both pest and predator were collected from a cotton ecosystem in Tirunelveli District, Tamilnadu, India then reared in a plastic container (15cm length, 10cm width and 5.5cm height), the lid of which had small holes for ventilation. Red cotton bug, *D. cingulatus* and *S. litura* life stages were collected from cotton fields, Tirunelveli District, Tamil Nadu and maintained on young, potted cotton plants under laboratory conditions. Life stages of these pests emerging in the laboratory were used for the experiments. The reduviid population was reared as described by Sahayaraj (2003) in a mixture of *D. cingulatus* and *S.litura*. The laboratory-born *R. marginatus and R. fuscipes* third, fourth, and fifth nymphal instars and adult predators were incubated at six constant temperatures. The following constant treatment levels were used in environmental chamber studies: 15, 20, 25, 30 and 35± 0.5°C under 60 ± 10% RH and a photoperiod of 13:11 (D:L). Third, fourth and fifth instar nymphs and adults of *R. fuscipes* and *R. marginatus* were maintained inside the environmental chamber and provided with a mixture of *D. cingulatus* and *S. litura,* continually for four days. Observation was done daily to monitor survival and exuviations of reduviids nymphs. The studies were conducted in two steps: first, preferred stages of the predators were determined and second, by using the preferred stages, biological control potential of the reduviids was observed. To examine the influence of temperature on reduviid predatory efficiency, approaching (including/searching, attacking), handling time, weight gain, and number of prey consumed were recorded.

3. 2. 1. Stage Preference

A stage preference study was conducted on all the life stages (except the first and second instar) of *R. marginatus* with different life stages of *D. cingulatus* and *S. litura* by choice experiment. To study the stage preference, *R. marginatus* third instar nymph was introduced in to a petri dish (14cm width and 9cm height) containing fresh cotton leaves and *D. cingulatus* second, third, fourth, fifth nymphal instars and adults (each two) were released separately and the predatory behaviour was observed visually for 4 consecutive hours. Successfully captured, killed and consumed prey stage was recorded as the preferred stage of the reduviid. Similar procedure was followed for other life stages of the predators. In another study *R. marginatus* and *R. fuscipes* were provided with all five larval stages of *S. litura* separately and the preferred pest stage was recorded. Fifteen replications were maintained for each life stage of the predator separately. Once stage preference was known, then the biological control potential evaluation studies were conducted using the following procedures.

3.2.2. Bioefficacy

Preferred stages of *Dysdercus cingulatus* (5 preys/ container) were introduced into the container containing cotton twig (*Gossyppium hirsuticum*) and it was allowed to acclimatize for 1hr. Then life stages of *R. marginatus* and *R. fuscipes* were introduced into the same container separately and the feeding events like approaching time, handling time, etc., were recorded continuously for 4 hours with visual observation. After 24 hours, weight gained and number of prey consumed by a predator was recorded and considered as predatory rate. Fifteen replications were maintained for each life stage of both predators separately. Similar procedure was also followed for *S. litura*.

3.3. OBSERVATIONS

3.3.1. Stage preference

Stage preference is the first step in the biological control evlaution studies of any natural enemies. *Rhynocoris marginatus* third, fourth and fifth nymphal instars and adults preferred second, fourth and fifth nymphal instars of *Dysdercus cingulatus* respectively. The preference was different when *R. marginatus* was provided with life stages of *S. litura* (second, third, four and fifth instar larva). *D. cingulatus* second, third, and fourth nymphal instars were preferred by third, fourth and fifth nymphal instars and adults of *R. fuscipes* respectively, whereas, third, fourth, fifth nymphal instars and adults of *R. marginatus* preferred second, fourth and fifth nymphal stages of the red cotton bug. In contrast, third, fourth, and fifth nymphal instars of both reduviids preferred second, third, and fourth instar larvae of *S. litrura*, respectively. Adults of *R. fuscipes* preferred only fourth instar larvae; however, *R. marginatus* adults capture and consume fifth instar larvae.

3.3.2. Bioefficacy of *Rhynocoris marginatus*

The bioefficacy of *R. marginatus* on *D. cingulatus* reveals that based upon the temperature variation both the predatory behaviour and bioefficacy varies. For instance at room temperature, *R. marginatus* third, fourth, fifth nymphal instars and adults consumed 3.33, 3.18, 2.44 and 3.93 *D. cingulatus* respectively. While we subjected the reduviids in different temperature regimes, the predatory rate was reduced (Figures 3a) except in fifth instar *R. marginatus* at 25 and 30°C. However, it was statistically insignificant (df = 4, P = 0.339; df = 7, P = 0.165) when compared with the control category (P = 0.05). Even though the predatory rate was lower than the control, the weight gain of *R. marginatus* (third, fourth and fifth nymphal instars) was increased at 30°C and also at 25°C in the fourth instar. Approaching and handling times gradually decreased and increased from 10 to 35 respectively. Then it was slightly decreased at 35°C. These predators preferred the ventral side of *D. cingulatus* and abdominal region on *S. litura* (Figure 3b).

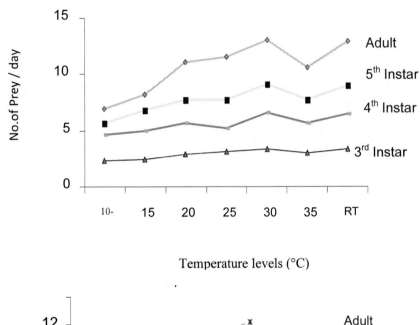

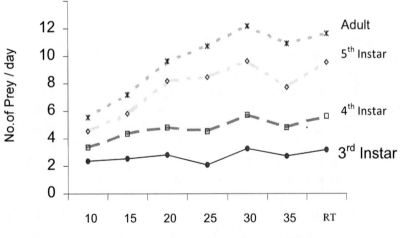

Figure 3. Predatory rate of *R. marginatus* life stages on *D. cingulatus* (a) and *S. litura* (b) at various constant temperatures (°C)

Another important parameter of concern in the bioefficacy was handling time. Handling time was gradually increased from 10 to 30°C and then decreased at 35°C. This concept was common in all the life stages of both the reduviids and their preys. Statistical comparison between control to 15°C of third instar, 20 and 30°C in fourth instar and 10 and 30, 35°C on proves insignificant (df-6; P=0.578; df = 7; P=0.572). Other temperature categories significantly influence handling and approaching time (df =6; P = 0.009) of the fifth instar (df = 7; P = 0.002). The handling time was gradually increased from third instar to fourth instar and then gradually decreased to fifth nymphal instar and also to adults when *D. cingulatus* was offered as a prey. While we offer *S. litura* as a prey, handling time was higher in fourth instar nymphs of *R. marginatus* followed by adults, third and fifth instars.

3.4. BIOEFFICACY OF *RHYNOCORIS FUSCIPES*

Handling time was gradually increased up to 25 and 30°C when *D. cingulatus* was provided with third and fifth nymphal instars and fourth nymphal instar and adults of *Rhynocoris fuscipes* respectively (Figure 4a). From the figure, it was vary clear those nymphal instars and adults of *R. fuscipes* took maximum time at 30°C and 35°C respectively to approach *S litura* nymphal instars and adults respectively. Similar trend was also observed for handling time too. However, irrespective of *R. fuscipes* life stages, it consumed maximum *D. cingulatus* at 30°C. Life stages of *R. marginatus* consumed maximum amount of *D. cingulatus* at 30°C except in fourth nymphal instars, especially in the fourth instar *R. marginatus*. All other life stages consumed maximum adults on *S. litura*

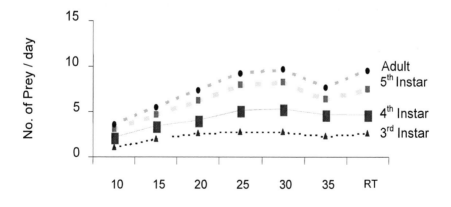

Temperature levels

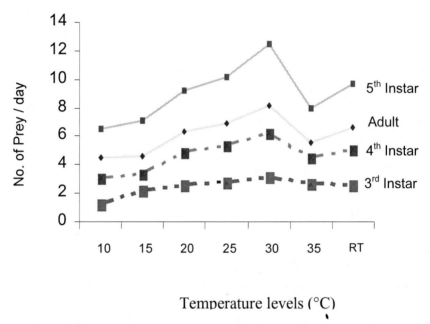

Temperature levels (°C)

Figure 4. Predatory rate of *R. fucipes* on *D. cingulatus* at various *D. cingulatus* (a) and *S. litura* (b) temperature levels (°C)

Weight gain ranged from 0.44 mg at 10°C to 2.41 mg at 30°C on third instar. When the temperature attained a moderate level weight gain was increased. But lower (10°C) and higher temperatures (15°C) showed minimum weight gain. Between the two cotton pests, the biocontrol efficiency of both reduviids was high on *S. litura*. Another finding reveals that of the six temperature regimes, *R. fuscipes* more actively feeds at 30°C. When *R. marginatus* was provided with *S. litura* third and fourth instars between 30 and 25°C, approaching time was significant (df = 4; P=0.016; df = 7; P=0.019) (Figure 4b). For the weight gain comparison between the six temperatures and control were highly significant by paired sample 't' test (df = 7, P= 0.000; df = 5, P=0.002; df=5, P=0.062; df = 7, P=0.025 for third, fourth and fifth nymphal instars and adults of *R. fuscipes* respectively on *D. cingulatus* at control). When *R. fuscipes* was fed with *S.litura,* predator weight gain [(P=0.000) for 10 to 20°C followed by 25 to 35°C df = 3, P=0.008 and df = 5,P=0.002; df = 7; P=0.008] and approaching time (P= < 0.082 and > 0.022) were statistically significant except for 25 to 30°C. Irrespective of the *R. marginatus* life stages, the predatory rate was higher at 30 °C for both pests. Among the life stages, the predatory rate was higher on adults followed by fifth, third, fourth and third nymphal instars of *R. marginatus*. As observed for *R.*

marginatus, the predatory rate of *R. fuscipes* was also higher at 30°C. However, the predatory rate of *R. fuscipes* fifth nymphal instar, and adult were differed on *D. cingulatus* to *S. litura.*

3.5. DISCUSSION

The results of the current study demonstrate the searching and handling ability of two reduviid predators and suggest their biological control potential against life stages of *S. litura* and *D. cingulatus.* Between the two pets tested, larvae of the cotton leaf worm *S. litura* were less aggressive than *D. cingulatus. D. cingulatus* easily escape from the predators by vigorous thrashing movements, so the predators delayed approaching *D. cingulatus.* Results reveal that temperatures have no influence on the prey stage preference. However, from the results, it was very clear the younger nymphal instars of these predators preferred younger (small) nymphal instars of *D. cingulatus* and larvae of *S. litura.* It was also recorded that, *R. fuscipes* and *R. marginatus* did not attempt to prey *D. cingulatus* and *S. litura* which are smaller when compared to their own body.

The genral predatory pattern of reduviids, which includes arousal, approach, capturing, rostral probing (capturing), injecting toxic saliva, paralyzing, sucking (handling), and post-predatory acts, has not been altered by the temperatures. The approaching time (or attack rate) and handling time were the parameters used to determine the magnitude of biological control efficiency and bioefficacy of any natural enemies. The attack rate and handling times differed significantly among various temperatures, indicating that the predatory reduviids respond differently to pests tested. *S. litura* larva too escaped quickly at moderate temperatures (25°C and 30°C), rather than the lower temperatures (10°C and 15°C). Generally reduviids captured their preys, which move much faster than their slow-moving preys. Similar observations were recorded in other reduviids (Haridass and Ananthakrishnan, 1981; Maran, 1999; Maran et al., 2002) and also in Pentatomid bugs (De Clercq and Degheele, 1992a; 1993). At 25 and 35°C *R. marginatus* and *R. fuscipes* were very aggressive and they approached immediately repressed and handled more prey, which in turn enhanced the prey consuming ability. The results imply that the predator will spend a large amount of time with non-searching activities (eg: resting or moving here and there without feeding) at low temperatures, while positive searching and preying activities would be expected at higher temperatures. In Tirunelveli district in particular and Tamil Nadu, India in general, both *D. cingulatus* and *S. litura* populations in cotton is maximum during

May to July (our personal observations), when temperature is usually < 30°C, and the population of both reduviids reach their annual peak.   Usually the *R. marginatus* population remains high at field land (Kumaraswamy, 1991). Therefore, a significant natural control effect of the reduviid on *D. cingulatus* and *S. litura* could be expected at that time.

A quantitative and qualitative understanding of the interaction among different species is crucial for the management of agricultural pests. Each species in an agricultural ecosystem is part of an often-complex community, where it has to interact with many other species. Our laboratory tests revealed that the biocontrol efficacy of both predators were often similar in terms of handling and approaching time, even though a few specific activities like weight gain varied in accordance with the temperatures (higher and lower). This result corroborated the observations recorded in generalist arthropod predators observed by Campbell et al. (1974) and Thompson, (1978). They postulated that predation of the generalist predator was mainly influenced by physical factors such as temperature. From the result it was evident that among the six temperatures, the bioefficacy of *R. marginatus* and *R. fuscipes* was high at 25°C and 30°C. Prey consumption generally increased as temperature increased (Anderson et al. 2001; Isman and Chapman, 2001; Jalali et al., 2010). Similar findings were also recorded by De Clercq and Degheele (1992a, 1992b). They demonstrated that temperature not only reduces the predatory activity but other factors like bioefficacy performance as mentioned above. Furthermore, temperature acts on insects as a stressor when it drops below or increases above the insect's threshold temperature range.

Successful capturing of prey would greatly depend on the relative size and strength of the prey. The predator also favoured the temperature variation as reported in pyrochorrid predators (Shrewsburry, 1996; Anderson et al., 2001; Lie et al., 2005) but it was not varied when the prey size was common (Sahayaraj, 1995a, 1995b, 2001, 2003). It is a well-known fact that the predator required more time to search for prey at low temperatures and it spent more time on non-searching activities. Predatory rate of *R. marginatus* gradually increased at 20°C (2.35) to 32.5°C (3.20) when *C. cephalonica* was provided as a prey (Morgan, 1985; Sahayaraj et al., 2004). Efficiency of prey capture of reduviids increased with increase in temperature, along with the number of prey consumed. At 30 and 35°C, the movements of predators as well as prey were faster than during other temperatures. The over all number of prey consumed, as also the quantity of prey consumed, was higher at 30 and 35°C than at 25°C. Lower food intake at lower temperatures might be due to the remaining of food in the foregut for longer periods as reported by Usha Rani (1992).

Several potential direct effects of temperature on predators (or) prey may observe prey vulnerability and exposure to predators (Anderson et al., 2001). Direct temperature effects may influence prey detection of natural enemies, alarm signaling, escape behaviours and / or defense (Gilchrist, 1995), as well as predator foraging (Morgan, 1985), prey handling (Thompson, 1978) and metabolism (Schultz et al., 1992). Early researchers Markkula and Roivaines (1961) suggested that altering abiotic factors, such as increasing light exposure and altering temperature factors by decreasing ambient moisture levels, might discourage natural enemies. From this point of view, we found out that temperature differences between environmental conditions and the experimental level mainly influenced the biological control efficiency, which was higher than at room temperature. A similar view has already been expressed in other predators by Shrewsburry (1996).

Rhynocoris fuscipes and *Rhynocoris marginatus* bioefficacy on *D. cingulatus* and *S. litura* were higher at 30°C. In India cotton has been cultivated in all the states where the temperature ranged from 15 to 35°C. The foraging behaviour of an insect is greatly determined by temperature. Moreover, reduviids' prey consumption has been found to increase as temperature increases. This indicates that these reduviids exhibited thermoregulatory behaviours or possessed broad or plastic operational temperature ranges and were capable of foraging in varying thermal environments. This kind of similar trend was previously recorded in *Chrysoperla carnea* (El − Walkil, 2003; Mahdian et al. 2006), *Picromerus bidens* and *Podisus maculiventris* (Heteroptera: Pentatomidae) (Mahdian et al., 2006).

Aging has been considered as a declining change from maturity to senescence and has been widely studied in insects, particularly reduviids (reviewed by Ambrose, 1999). Sahayaraj and Ambrose (1996); Sahayaraj (1995, 2004; Ambrose and Sahayaraj (1996) reported that a linear relationship existed between reduviid predator age and bioefficacy whereas Sahayaraj (1994), Sahayaraj and Ambrose (1994) suggested that prey age influenced reduviid bioefficacy. Factors such as large searching areas, host plants, and weather under field conditions may influence the effectiveness of predators. But in this study, we placed cotton leaves in the experimental arena and maintained a constant temperature. Among the nymphal stages tested, fifth instar nymphs invariably took more time for consuming and subsequently sucked more amount of prey. It leads to more increased weight gain than in adults.

Studies are needed for evaluating more reduviids because the availability of additional reduviid predators of *D. cingulatus* and *S. litura* would prove to be an asset in the successful biological control of these pests under various situations. Apart from natural conditions many factors are known to influence the biological control potential of predators (Islam and Chapman, 2001; Sahayaraj et al., 2004). This can be attributed to the fact that the predators are more active and have greater predatory potential rates at higher temperatures. This implies that the predators spend a large amount of time with non-searching activities (resting) at low temperatures, while positive searching and the preying activity could be expected at higher temperatures. Among these, temperatures have a profound influence, as govern the rate of growth and development and prey consumption (Shrewsburry, 1996; Anderson et al., 2001; Islam and Chapman, 2001). There has been no systematic evaluation of the influence of temperature on prey consumption by reduviid predators. Such information could be useful for predicting the potential of these predators under varying environmental conditions. The current study shows that reduviids display relatively high predation rates on *D. cingulatus* and *S. litura* at a wide range of temperatures, indicating its potential for augmentative releases against these pests. Furthermore, knowledge of the relation to temperature and prey in reduviid predators might be useful for deciding the release of this predator in natural environments.

Chapter 4

EGG AND ADULT MACROMOLECULES

ABSTRACT

As a framework within which to describe the function of the main biochemical compounds involved in an insect-rearing environment, the desiccation and the daily resetting of critical thermal thresholds affecting mortality and mobility have emphasized the role of temperature as the most important abiotic factor, acting through physiological processes to determine biochemical constituents. From the results it is very clear that, the total carbohydrate content is low at 10°C (22.2 μg/mg). It gradually increases when the temperature is increased and reached its peak at 35°C. In another study biochemical composition of eggs in relation to temperature was observed. Regarding the total carbohydrate, maximum content was recorded at 35°C for the two reduviid species. Furthermore, egg protein contents gradually increased up to 30°C in both the reduviid species gut protein profile of *R. fuscipes*. Biochemical composition of eggs in relation to temperature noted the total carbohydrate; maximum content was recorded at 35°C on both the reduviid species. Furthermore, egg protein contents gradually increased up to 30°C in both the reduviid species.

4.1. INTRODUCTION

Temperature can induce behavioral modifications, such as changes in locomotion activity or flight ability, as well as a range of biochemical and physiological responses (Hochachka and Somero, 2002). Particularly well-known is the effect of extreme temperatures on cellular responses, such as the expression of protective heat shock proteins at high temperatures (Bahrndorff et al., 2009) or late embryogenesis abundant proteins when exposed to freezing conditions

(Thomashow, 1998). However, even a moderate change in temperature can lead to vital physiological and biochemical adjustments, one of which is a change in fatty acid composition.

The environmental temperature on insects causes dramatic changes in behavioural, physiological activities and biochemistry (Agrell and Landquist, 1973; Dooremalen and Ellers, 2010), particularly the haemolymph lipoprotein in a haematophagous reduviid, *Triatoma infestens* (Maria *et al.*, 1991; Rolf *et al.*, 1999). Furthermore, Jeffrey and Jesusa (2006) assessed the biochemical fitness of the predator of *Podisus maculiventris* in relation to food quality and effect of five preys. The information on the influence of temperature on various biochemical entities was available in the literature (Maa, 1987). Proteins constitute the basic entities in the living organisms and undergo both qualitative and quantitative changes during development (Engelman, 1979). The fat body is the principal storehouse of lipids in insects. Most of the lipid is present as a triglycerol which commonly constitutes more than 70% of the dry weight of the insect fat body (Chapman, 1998; George and Ambrose, 1999b). Lipids are synthesized from the fat body and secreted into haemolymph through physical activities (Brooks, 1969; Beenkkers *et al.*, 1985). Its importance along with protein is also available for the literature of heteropteran insects (Kunkel and Nordin, 1985; Helosia et al., 1997). Further more Beenkkers et al. (1985) viewed that the whole body content is an important carbohydrate reserve in many insects. We evaluated the impact of constant temperature on the whole body and egg macromoleular profile of these two reduviids.

In field studies, the mean number of prey items in a generalist predator gut may be as a few separate PCR assays evaluated (Gozlan et al., 1997; Harper *et al.*, 2005). This technique is effectively peculiar for many useful field-based ecological studies. Rapid PCR-based screening systems needed for the study of the prey diversity of generalist predators have been developed to expend the potential of molecular detection in to various areas of food web research (Dodd, 2004). From the published results, it becomes very clear that except for the haematophagous reduviids such as *Trypanosoma rangeli* and *Trypanosoma cruzi* (Moser *et al.*, 1989; Breniere *et al.*, 1999; Russomando *et al.*, 1996; Shiankal *et al.*, 1996; Vallejo *et al.*, 1999), to date, no information is available for the polyphagous reduviid predators. Ehrlich developed the Polymerase Chain Reaction (PCR) technique in 1989. It is one of the simplest, fastest and least expensive molecular approaches to use. RAPD-PCR (Randomly Amplified polymorphic DNA) (Shappiro *et al.,* 1988) is used to amplify a region of DNA that lies between two regions of known sequence (Teresa *et al.*, 2002). PCR method has been used in many fields including understanding the genetic variability in insects (review of Sheppard and Hardwood, 2005).

Therefore, it is envisaged to analyze quantitatively and electrophoretically in relation to temperature modification also on imperative one. This chapter deals with changes of whole body and egg macromolecules (carbohydrate, protein and lipids) of two reduviid predator species by spectrophotometry whole gut protein by electrophoresis method; gut DNA polymorphism by RAPD – PCR analysis by AGE in relation to six different constant temperatures on *R. marginatus* and *R. fuscipes*.

4.2. METHODOLOGY FOLLOWED

Freshly laid eggs were incubated for 12 to 15 days (10, 15, 20°C), 6 to 7 days (25 to 30°C) and 4 days (35°C); they were used for analyzing protein, carbohydrates and lipids. Thirty laboratory-emerged adult reduviids of both male and female sexes were maintained at 10, 15, 20, 25, 30 and 35°C separately until the start of the experiment in the BOD incubator. After one month, ten insects were randomly selected and their whole body total carbohydrate (Sadasivam and Manickam, 1997), protein (Lowery *et al.,* 1951) and lipids (Bragdon, 1951) were estimated using glucose, bovine serum albumin and cholesterol as standards, respectively.

4.3. OBSERVATIONS

4.3.1. Eggs macromolecules profile

The biochemical composition of eggs in relation to temperature was evaluated. Regarding the total carbohydrate, maximum content was recorded at 35°C for the two reduviid species. Furthermore, egg protein contents gradually increased up to 30°C in both the reduviid species. In contrast, lipid content gradually decreased up to 35°C in *R. marginatus* and *R. fuscipes* (Figure 5).

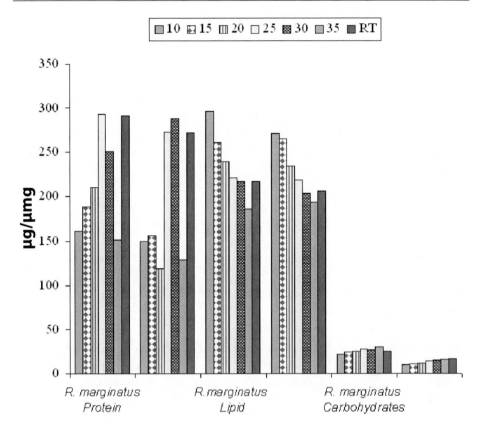

Figure 5. Fate of total protein, carbohydrate and lipid (in g/100mg) of *R. marginatus* and *R. fuscipes* eggs at various constant temperatures ($^{\circ}$C)

4.3.2. Whole animal macromolecules profile

Total carbohydrate, protein and lipid content of *R. marginatus* in relation to different temperature regimes (10 to 35°C) is presented in figure 5. From the figure it is very clear that, total carbohydrate content was low at 10°C (22.2 µg/mg). It gradually increased when the temperature was increased (23.5, 25.11, 27.8 and 28.8, µg/mg for 15, 20, 25 and 30°C, respectively) and attained its peak at 35°C (30.14µg/mg). Statistical analysis between controls (25.3µg/mg) with different temperatures showed that levels namely 25, 30 and 35 were significant at 5% level. Similar observation was also recorded in *R. fuscipes* too. Between the two reduviid species, *R. fuscipes* had more carbohydrate content than *R. marginatus*.

In contrast to the carbohydrate, lipid content decreased gradually from the lower temperature to the high temperature levels (Figure 6). From the fate of body biochemical content, results showing the carbohydrate content were almost low and lipid content was high in *R.marginatus* and *R. fuscipes* respectively. In the case of protein content, it was clearly observed that the control category (291.20 μg/mg), 20°C (290.71 μg/mg) and 25°C (293. 54μg/mg) were more or less similar. Consequently, the protein level in control (272.0 μg/mg) and 25°C (273.28 μg/mg) categories were also noted as similar in *Rhynocpris fuscipes*. Statistical analyses were made between the control and 10 to 30°C revealed that all the comparisons were significant at 5% level. Protein content increased gradually from 10 to 25°C and 10 to 30°C for *R. marginatus* and *R. fuscipes,* respectively.

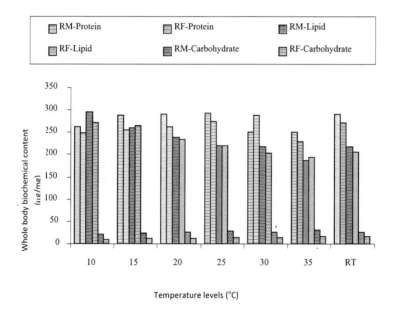

Figure 6. Influence of constant temperatures (°C) on whole body macromolecular composition (in μg/mg) of *R. marginatus* and *R. fuscipes*

4.4. DISCUSSION

Temperature is an important environmental factor, which affects the growth and development of all living organisms including insects. Growth of an insect below optimum temperature may bring about physiological and quantitative changes in metabolic activities. It is a general observation by the entomologists that lowering of growth temperature resulted in increased production of pigments, synthesis of polysaccharides, and increased levels of unsaturated fatty acids in cellular lipids. From the second chapter, it is very clear that different reduviid predators can tolerate temperatue differences within certain limits. This tolerance is provided through modifying metabolic activities. Therefore, it may result in changes in total carbodydrate, protein and lipid levels. Studies done with natural and artificial foods have shown that carbohydrate, protein, and lipids were the main foods in the various physiological activities of reduviid predators (Sahayaraj, and Ambrose, 1994a; Sahayaraj et al., 2004).

Appearance of some protein substance was mainly responsible for temperature variation as expressed by Bradford (1976). According to Ryan and Dick (2001), whenever temperature reached a peak (or) higher (or) exceeded $35^{\circ}C$, adult insects were unable to ultimately adapt; also the changes that were caused were mainly based upon the treated temperature. This kind of opinion was also brought out by Shappiro et al. (1988). Temperature changes also affected the Lepidopteran pests; it may be related to heat shock proteins induced by sub-lethal temperature and other environmental stresses (Dodd, 2004; Zhang et al., 2007; Reza et al., 2008). According to the report, major heat shock causes them to disintegrate at different temperature levels from 33 - $41^{\circ}C$ after 2 to 3 hr at $20^{\circ}C$. Furthermore, the energy as well as break down of food particles of insects and the metabolic process is directly or indirectly controlled by biotic factors as suggested by Salt (1965). From the results, it is very clear that a moderate change in ambient temperature can lead to vital physiological and biochemical adjustments in ectotherms, one of which is a change in total carbohydrate, protein and lipid compositions.

Eggs are small, and therefore nearly always isothermal with nearby microenvironments; embryo temperature is determined by maternally chosen microhabitats. Here, we focus on predatory insects that oviposit, as most do, directly onto soil. For these eggs, microsite experience reflects interactions between ambient environmental characteristics (e.g. air temperature, solar radiation, wind speed) and soil morphology. Eggs lie in boundary layers of relatively still air, which resist heat and moisture transfer between a leaf and its surroundings. Macromolecules like carbohydrate, protein, and lipid content of *R. marginatus* and *R. fuscipes* had been dear in the lifetime. Previously George and Ambrose (1999a, 1999b); George *et al.* (2002); Ambrose and Maran (1999b; 2000a, 2000b) demonstrated that insecticides affected the carbohydrate, protein, and lipid contents of *R. marginatus*.

Among the three macromolecules mentioned initially the whole body carbohydrate content could be attributed to its higher utilization; also the energy releasing site warranted altering metabolism due to an agreement observation discussed in other insects belonging to Coleoptera (Pigman and Horton, 1970). Such high-energy demands for various endothermic biochemical reactions could readily react on carbohydrate reserves because they are the principal and immediate source of energy precursors (Wyatt, 1976). Moreover, prolonged treatment of cooling as well as higher temperature affected function either quantitatively or qualitatively at the range of metabolical canges. It led to reduce the synthesis of protein by rearranging the protein synthetic machinery.

Chapter 5

GUT MICROBIAL PROFILE AND THEIR HYDROLYTIC ENZYMES PRODUCTION

ABSTRACT

The indigenous gut bacteria are regarded as a valuable metabolic resource to the nutrition of the hosts by improving the ability to live on sub-optimal diets and digestion efficiency, acquisition of digestive enzymes and provision of vitamins. Thirteen and twelve bacteria were identified based on morphological characters, and biochemical tests in both *R. fuscipes* and *R. marginatus*. The bacterial species isolated from the gut of *R. marginatus* included, *Staphylococcus aureus, Bacillus cereus, B. magaterium, Enterobacter aerogenes, Micrococcus luteus, Corynbacterium kutcherii, C. xerosis, Bacillus subtilis, Escheritia coli, Pseudomonas aeruginosa* and *Micrococcus variance*. Protease activity was high at 25°C in the fore and hindgut of both predators whereas amylase and invertase activity was highest at 35°C. In *Rhynocoris marginatus*, foregut and hindgut showed maximum esterase activity at 20°C. But in *Rhynocoris fuscipes*, esterase activity was maximum and equal both at 30°C and room temperatures. All the bacterial species were positively correlated such as 0.67, 0.50, 0.84 and 0.42 for *B. cereus, M. variance, S. aureous* and *C. xerosis* respectively, accompanied with each different temperature except *B. subtilis* and *Lactobacillus delbrucki*, which showed negative correlation to temperatures. It was more predominant (29%) at 10°C. In general *S. aureus* and *B. cereus* were more or less similar in population at all temperatures in both the reduviid species. These two bacteria were considered as autochthonous bacteria of reduviid predators studied.

5.1. INTRODUCTION

The effect of climate on organisms, communities, and the environment at large has become a pressing issue for biologists and environmental scientists. It is important to note that many macroorganisms including predartory insects, live in symbiosis with microbes and that host fitness may be affected indirectly by higher temperatures due to the disruption of mutualistic relationships. Despite the current interest in insect-microbe symbioses, the vast majority of such systems have been poorly studied. A group of insects that has recently received some attention are the true bugs (Hemiptera, Pentatomomorpha). Studies in the early 1900s suggested that mutualistic bacteria colonized a portion of the gut of insects in different pentatomomorphan families. More specifically, monocultures of bacteria were present in high densities in the crypt- (or cecum-) bearing organ preceding the hindgut of hosts, with different bacterial taxa associated with different true bug families. Pentatomid symbionts are polyphyletic and closely related to plant-associated bacteria in the genera *Erwinia* and *Pantoea* (Prado and Almeida, 2009). However, very little information is available for reduviids, another true bug family. Therefore, it is plausible that observed effects of climate on species distribution or performance might stem from disruption of symbiotic interactions as much as from direct effects on host biology. Despite the current interest in insect-microbe symbioses, the vast majority of such systems have been poorly studied. A group of insects that has recently received some attention are the true bugs (Hemiptera, Pentatomomorpha) (says Prado et al. 2010). Previously Prado et al. (2009) reorted that the prevalence of the gut symbiont was reduced or it was absent in second-generation insects of both species, results similar to those for another pentatomid, *Nezara viridula* (L.) suggesting that temperature has an important role in the maintenance of gut symbionts in this insect family.

Enzymes are proteins, which catalyze a variety of reactions in the biological systems. The multifarious enzymes evident in living cells can be isolated from the various biological active sites. Different techniques have been used to measure the enzyme of the salivary glands using standard histochemical and calorimetric methods for proteolytic, macerating and cellulolytic enzymes (Balogun, 1972; Barnard and Prosser, 1973). PIs inhibit the gut proteinase of the insect, which adversely affects the protein digestion in the gut and forces the insect to synthesize alternative proteases to compensate for the inhibited activity. The insect causes severe quantitative and qualitative damage to crops by feeding on leaves, stems and grains. Its feeding is typical of heteropteran, piercing and cutting tissues with their stylets while injecting digestive enzymes, amylases and proteases through the salivary canal to liquefy food into nutrient-rich slurry. The food slurry is ingested through the food canal and passed into the alimentary canal where it is further digested and absorbed (Saxena and Bhatnagar, 1961; Biggs and Greego, 1994). In addition to the enzymes, precursors like free amino acids mainly fluctuated by pH, salt concentration and temperature, in general are also present in the literature. The digestive enzymes commonly found in the salivary secretions and regions of the digestive tract of various insects have been examined by many authors and were comprehensively reviewed by House (1965). Regional localization of various enzymes in the alimentary canal of *Sophrorhinus insphsbus* (Coleoptera) (Adedire and Balorgun, 1995) and Reduviids (Cohen, 1993) has been recorded earlier.

Microorganisms play an essential role in the growth and development of many insects. Numerous investigations available on the bacterial flora of insect guts reveal that the bacterial flora appears to be a fortuitous contamination of the insect's surroundings and its food (Hunt and Charnley, 1981; Brauman et al., 2000, 2001). The indigenous gut bacteria are regarded as a valuable metabolic resource to the nutrition of the hosts by improving the ability to live on sub-optimal diets and digestion efficiency, acquisition of digestive enzymes and provision of vitamins (Bignell et al., 1997; Chen and Purcell, 1997). Many studies have reported the number of intestinal microflora, which got closely associated with the feeding habits of insects (Jones, 1984).

Very little information is known about the autochthonous gut bacterial populations associated with the reduviids (Sahayaraj and Mary, 2003). Recently, Sahayaraj (2007b) isolated and identified the gut bacteria of three reduviid predators such as *Acanthaspis pedestris* (Stal.), *Haematorrhophus nigrovidaceous* (Reuter) and *Cattamierus brevipennis* (Distant) for the first time. However, no information was available on the impact of temperature level on the gut autochthonous bacterial population and their impact on hydrolytic enzyme activities of *R. marginatus* and *R. fuscipes*.

5.2. EXPERIMENTATION

Laboratory-emerged *R. marginatus* and *R. fuscipes* adults were maintained at different temperatures in the study. The stock categories were cultured under laboratory conditions (29 ± 1°C and 80% RH) on *C. cephalonica* following the methods of Sahayaraj (2002).

5.2.1. Preparation of gut sample

Ten each of *R. marginatus* and *R. fuscipes* were selected randomly from all the temperature regimes prior to the (morning) feeding when the gut was empty and the insects were placed at 4°C for 05 minutes prior to use. Surface of the predators was sterilized with 0.1% Mercuric Chloride for 2 minutes and washed with sterile distilled water thrice for washing away the external impurities present on the experimental predators. Under aseptic conditions each insect was carefully dissected using sterile pins, fine forceps and razors in a dissection tray filled with sterile phosphate buffered saline (PBS) (pH 6.9). Guts were isolated individually, washed several times with fresh phosphate buffered saline to minimize the possible microbial contamination and used for the study. Wet weight of the gut was recorded for individually categorized with six temperatures for *R. marginatus* and *R. fuscipes*.

5.2.2. Enumeration of THMP of predators' gut content

The isolated gut was homogenized under aseptic conditions with sterile insect ringer solution (IRB) in mortar and pestle. The homogenate was filtered through Whatmann filter paper No.1 and the pH was measured using pH meter. The filtrate was serially diluted in sterile saline and 0.1 ml of aliquot was plated on nutrient agar (NA) and Trypticase soy agar (TSA). The seeded nutrient agar plate (NA) was inoculated at 37°C for 24-48 hours whereas; the Serially Dilution Agar (SDA) plates were incubated in 28°C for 48-72 hrs. Microbial colonies appeared after the incubation period was enumerated and the numbers of colony forming units were expressed as a wet weight of the gut.

5.2.3. Identification of Microbes

Different morphological microbial colonies were selected, sub-cultured and stored at 4°C on respective agar slants. Bacterial strains were identified using the criteria suggested by Cappucino and Sherman (1999) based on morphological, cultural and biochemical characterizations.

5.2.4. Hydrolytic extra cellular enzyme

The extra cellular enzymes like amylase, protease, cellulase and gelatinase activities were tested by using the nutrient media containing 0.2% (W/V) carboxymethyl cellulose (cellulase), starch (amylase) and skimmed with powder (protease) as substrates. Pure cultures of each bacterial isolate were streaked on respective media and utilization of the substrates was determined by observing the clear zone around the colonies (Buchanon and Gibbons, 1979). All the screening experiments were replicated thrice.

5.3. OBSERVATIONS

5.3.1. Enzyme production

Protease activity was noticed both in foregut and hindgut, but the enzyme concentration was higher in hindgut than foregut. When the temperature was increased, protease activity was also increased from 10–25°C (0.216 to 0.591 μg/ml) in *R. marginatus*. Peak protease activity was recorded at optimum temperature 25°C, and then it got decreased when observed at 30 and 35°C. As observed in the foregut, the protease activity was also highest at 25°C (1.047) followed by 30°C (0.951 μg/ml) and at 35°C (0.946 μg/ml) (Figure 7). Similarly in *R. fuscipes,* maximum protease activity was recorded at 25°C [0.87 and 1.49 μg/ml for fore and hindgut respectively (Figure 8)].

Esterase activity was high both in fore and hindgut of *R. marginatus* at lower temperatures (20°C) (Figure 7). Then, the esterase level decreased when the temperature was increased. In *R. fuscipes* foregut, the esterase activity was almost high and equal values have been observed on both at 20 and 30°C. But in the hindgut activity was high during the temperature 30°C. This enzyme level varied slightly between fore and hindgut of these reduviid species.

Figures 7 and 8 show the amylase and invertase activities of *R. marginatus* and *R. fuscipes*, respectively. Irrespective of the predator species and location of gut, in general, activities of both invertase and amylase gradually increased when the temperature was raised from 0°C to 35°C. When the locations were compared, these enzymes activities were more pronounced in the hindgut than the foregut.

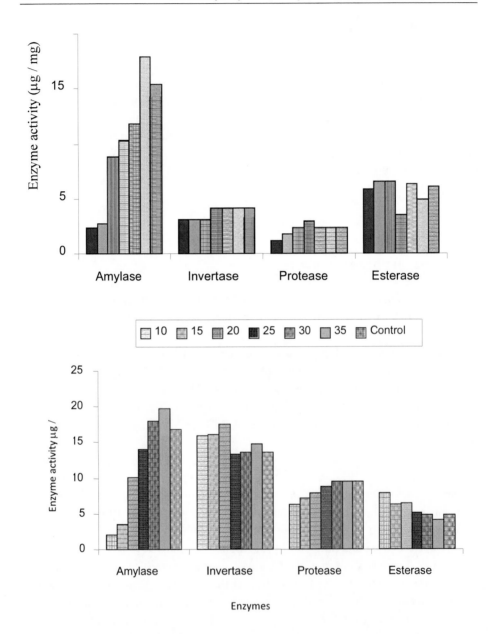

Figure 7. Influence of constant temperatures ($^{\circ}$C) on enzyme activity (in μg/mg) of *R. marginatus* in foregut (a) and hindgut (b) enzyme activity (in μg/mg)

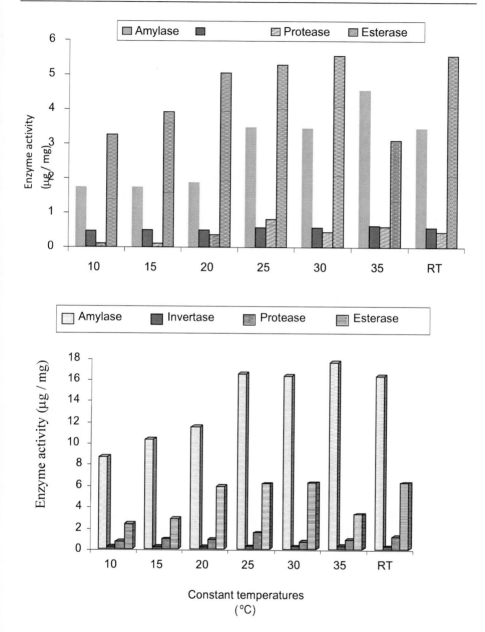

Figure 8. Influence of constant temperatures (°C) on *R. fuscipes* foregut (a) and hindgut (b) enzyme activity (in µg/mg)

5.3.2. Weight and pH of the reduviid guts

Generally, predator's intestinal pH profile was alkaline in nature (Lemeke, 2003). But the present study revealed that gut contents of *R. marginatus* (pH 7.6) and *R. fuscipes* (pH 7.0) were found to be either slightly alkaline or neutral. In *R. fuscipes,* gut weight diminished gradually up to 30°C then increased to 35°C whereas in *R. fuscipes* gut weight was gradually decreased up to 25°C then decreased from 30 to 35°C. Alimentary canal weight of *R. marginatus* was higher than that in *R. fuscipes.* Total heteroptrophic bacterial populations (THBP) of the whole gut homogenate of both predators. The THBP of both reduviid predators were gradually increased from 10-30°C and then declined at 35°C. Between the two reduviids, maximum THMP was observed in *R. fuscipes* rather than in *R. marginatus.*

5.3.4. Bacterial Composition

Thirteen and twelve isolates of bacteria were grouped based on morphological characters, and biochemical tests detected in both *R. fuscipes* and *R. marginatus* respectively. The bacterial species isolated from the gut of *R. marginatus* included *Staphylococcus aureus* (Rosenbach), *Bacillus cereus* (Frankland), *B. agaterium* (Carolus and Linn.), *Enterobacter aerogenes* (Rahn), *Micrococcus luteus* (Schroeter), *Corynebacterium kutcherii* (Kruse), *C. xerosis* (Lehman and Neumann), *Bacillus subtilis* (Ehrenberg), *Escherichia coli* (Migula), *Pseudomonas aeruginosa* (Migula) *and Micrococcus variance* (Pribram) (Table 10). Among them, *Staphylococcus aureus* (Rosenbach) was found to be the most dominant species both at 10°C and 15°C and *Micrococcus variance* was the dominant bacteria between 20 to 30°C. *Bacillus subtilis* (Ehrenberg) dominantly present at 15°C and 35°C for *R. marginatus* and *R. fuscipes. Micrococcus luteus* (Schroeter) and *C. xerosis* (Lehman and Neumann) were represented in low percentages at 25 and 20°C respectively. Among the observed bacterial species, *Escherichia coli* were observed only at 30°C. At an average of all the temperatures, *M. variance* constituted the dominant bacterium (40.56%) followed by *S. aureus* (31.78%) and *B. cereus* (24.10%). The most predominant bacterium observed in *R. fuscipes* (Table 8) was *M. variance* (47.22%) followed by *S. aureus* (30.50%) and *B. subtilis* (30.10%). All the bacterial species were positively correlated (0.67, 0.50, 0.84 and 0.42 for *B. cerosus, M. variance, S. aureus* and *C. xerosis* respectively) accompanied with each different temperature, except *B. subtilis* ($r^2 = - 0.35$) and *Lactobacillus delbrucki* ($r^2 = - 0.02$), which both show negative correlation to temperatures. In general, *S. aureus* and *B. cereus* were more or less similar in population at all the temperatures in both the reduviid species. In spite of both, low and high temperatures altered the bacterial species and their population level, although *Lactobacillus dellbruckii* (Orla - Jensen) and *Lactobacillus casei* (Hensen and Lessel) were present only in *R. fuscipes* (Table 11).

Table 10. Bacterial species composition (in %) of adults of *R. marginatus* subjected to various constant temperatures (°C)

Bacterial species	Temperature levels (°C)							
	10	15	20	25	30	35	RT	Mean
Bacillus cereus	37.26	31.2	15.39	12.21	0	0	0	24.03
B. megaterium	0	0	10.2	0	0	0	0	10.2
B. subtilis	0	21.0	10.01	10.0	14.83	25.07	14.83	15.95
Corynebacterium kutcherii	0	0	0	9.45	0	0	0	9.45
C. xerosis	0	0	17.21	0	2.18	2.47	2.18	6.01
Enterobacter aerogenes	13.72	0	11.25	15.36	28.36	0	0	17.2
Escherichia coli	0	0	0	0	1.63	0	0	1.63
Micrococcus variance	0	0	25.59	32.28	52.70	40.00	52.70	40.68
Micrococcus luteus	0	0	0	3.68	0	0	0	3.68
P. aeruginosa	0	0	0	0	0.55	6.34	0.55	2.8
Staphylococcus aureus	49.0	47.3	23.18	20.32	28.09	26.36	28.09	31.75

5.3.5. Hydrolytic extracellular enzymatic activity

Hydrolytic activity was observed in bacterial isolates of the whole gut in both the predator species. Of the four hydrolytic enzymes, cellulase activity was almost lower in these predators than the amylase and protease. Amylase activity was apparently higher at higher temperatures for *R. marginatus* (65.91) and *R. fuscipes* (67.6). Another hydrolytic enzymes protease showed peak activity at 25°C (55.1 and 47.8 for *R. marginatus* in *R. fuscipes,* respectively).

Table 11. Bacterial species composition (in %) of adults of *R. fuscipes* subjected to various constant temperatures (°C)

Bacterial species	Temperature levels (°C)							
	10	15	20	25	30	35	RT	Mean
Bacillus subtilis	0	44.15	23.17	0	0	23.07	0	30.1
B. cereus	41.20	10.21	17.30	16.05	0	0	0	21.2
B. megaterium	0	0	19.26	27.14	0	0	0	23.20
Corynbacterium kutcherii	0	0	0	4.39	0	0	0	4.39
C. xerosis	0	0	10.33	8.16	9.10	5.13	9.10	8.8
Enterobacter aerogenes	18.11	0	23.04	28.13	31.07	0	0	25.08
K. pneumoniae	0	0	3.21	0	0	0	0	3.21
Lactobacillus dellbrueckii	29.0	0	0	0	0.91	0	0.91	10.8
L. casei	0	11.60	0	0	0	0	0	11.60
Pseudomonas aeruginosa	0	0	0	0	0.71	2.0	0.71	1.14
Micrococcus variance	0	0	40.13	48.98	53.97	40.09	53.97	47.42
M. luteus	0	0	0	2.39	0	0	0	2.39
Staphylococcus aureus	55.1	44.15	15.99	15.08	22.7	36.92	22.7	30.37

5.4. DISCUSSION

An animal's digestive tract and its associated organs can account for up to 40% of the animal's metabolic rate, and because of these costs, the digestive tract must be tightly regulated in relation to food intake and quality. Thus, an animal's gut capacity, in terms of size and physiology, should maximize digestion from its natural diet; that is, dietary specialization should exist not only on the level of the community but also on the levels of gut size and digestive physiology. Gut microbial communities are rather ubiquitous, both in vertebrates and invertebrates. Symbioses between host and microorganisms range from pathogenic to mutualistic, facultative to obligate relationships. These gut-associated microbial communities play a major role in the metabolism of the host, in particular in the case of low and high level temperatures.

Digestive enzymes play a major role in the physiology of insects by converting complex food materials in to micromolecules necessary to provide energy and metabolites in *Sophrorhinus insphsbus* (Adedire, 1984). Amylase, protease, invertase and esterase showed maximum activity in salivary and haemolymph proteins of many insects (Chapman, 2000). Amylase is one of the key enzymes involved in digestion and carbohydrate metabolism in insects. Furthermore, Chatterjee et al. (1989) reported the presence and two different forms of amylase in digestive fluid and haemolymph. Abraham et al. (1992) noticed that amylase activity of the digestive fluid was 40 fold higher than that of haemolymph.

The major function of the digestive enzymes in the reduviid is "extra-oral digestion". Enzymes were mainly used to disintegrate prey tissue before ingestion after which further digestion took place. Cohen (1998) called this type of digestion as enzymatic tissue maceration was observed in *Zelus renardii*. While feeding, the reduviids not only feed on the haemolymph but also the interior contents including the organs (cells, tissues) and their networking macro and micro molecular complex including proteoglycans, collagens, elastics, etc. The nutrient rich materials in the prey are packed in a basement membrane that is impermeable of digestive enzymes (Agusti and Cohen, 2000; Lisa and Diane, 2009). The enzymes like trypsin and chymotrypsin are present in the reduviids (Cohen, 1998) used for digesting these materials too.

Salkeld (1961 and 1965) reported cathepsin the proteinase in the posterior midgut of *Sinca* spp. and *Zelus renardii* could liquefy and extract all of the nutrients of a prey nearly equal to its own body weight in less than two hours (Matsumara, 1988). Both in *R. marginatus* and *R. fuscipes,* the protease level in hindgut and foregut was increased. Although quite a good number of reports exist on esterase patterns in insect tissues, no information is available on the reduviid predators. Esterase mainly plays a lipolyticrole in eggs. This study shows the importance of esterase in digestion too.

The enzymic activities of the hindgut were higher than those of the foregut. Moreover, food protein stimulates the secretion of more protease in hindgut (Ishaaya, *et al.,* 1971, Upadhyay and Misra, 1991 and 1994). This study shows that in addition to the abiotic factors like prey (Sahayaraj, 2007a; Sahayaraj *et al.,* 2007a), temperature also influences the production of protease in *R. marginatus* and *R. fuscipes.* Carbohydrates ingested by heterotrophic organisms undergo several metabolic steps, in the first of which polymorphic carbohydrates are cleaved into their monomers, which can pass through membranes. Invertase, thus, appears to be particularly important an enzyme for the insects. Given this general importance, surprisingly few studies have tried to quantify invertase activity in reduviids (Sahayaraj *et al.,* 2007; Cohen, 1993). Invertase usually is quantified *via* the release of glucose from sucrose. The effect of high and low temperatures may reflect a reduced enzymatic level. The activities of the organism are influenced either directly or indirectly by the environment. The extreme (too low or too high) conditions of the environment may upset the physiological aspect of the insects. Enzyme is one of the most important determinants of physiological characters. Bacterial communities are known to play important roles in insect life histories, yet their consistency or variation across populations is poorly understood (Aaron *et al.*, 2010).

The present study suggests that factors other than prey nutrients, such as temperature, are also involved in the production of enzymes from both fore and hindgut of the reduviids. The higher amylase and invertase activities recorded at 35°C in the hindgut homogenate were expected since the predator shows a greater amount of carbohydrate as reservoir, which is considered as an important metabolic food for insects (Eguchi, 1983). These enzymatic variations in the gut could also be due to the presence of the microbes in the gut. Hence screening of microbes is essential to know about the enzymes secreted in the gut. Previously Kalaiselvam and Arulpandi (2006) reported different kinds of protease enzyme but only the least number of proteases has been recognized in the digestive action and regulates the physiological process. These variations might be due to the activity of various autochthonous microbes present in the alimentary canal. Basic information about the bacterial flora of gut is highlighted in the next chapter.

The alimentary canal of insects provides a suitable substratum for the development of microorganisms because of concentrated nutrients and extended surface of the intestinal lumen. On the other hand, these associated microbes may play an important role in the digestion, nutrition and defense system of the host animals (Pankaj *et al.,* 2003). Little is known about bacteria associated with the Reduviidae, the large group of mostly zoophagous insects comprising the haematophagous and predatory insects. In the present study we identified many interesting bacterial species, which belong to sub divisions of the Proteobacteria. Stingbug, *Megalopta punctatissima* also possess this type of bacteria in its gut (Fukatsu and Hosokawa, 2002). The bacterial genera found in *R. marginatus* and *R. fuscipes* were *Streptococcus* spp., *Staphylococcus* spp., and *Micrococcus* spp. In insects, there are typical bacterial colonies found in the intestines of all the polyphagous and phytophagous insects (Tanada and Kaya 1993; Egert *et al.*, 2003). The finding of the present report also indicates the diversity of bacteria present in the whole alimentary canal of *R. marginatus* and *R. fuscipes* based upon this previously evaluated kind of similar results with various diet effects by Sahayaraj and Mary (2003). Thermostable enzymes can be obtained from both mesophilic and thermophilic organisms; thermophiles represent an obvious source of thermostable enzymes. Thermostable enzymes, which have been isolated mainly from thermophillic organisms, have found a number of commercial applications because of their overall inherent stability (Santo *et al.*, 1998). It was also reported that abiotic factors alter the gut bacteria populations (Tsuchida *et al.*, 2002). It may be hypothesized that the bacterial flora degraded some acid metabolites, which might have induced the pH gut of the lepidopteran moth, *Lymantria dispar* (Linn.) that presented as slightly alkaline nature (Broderick *et al.*, 2004).

Chen and Purcell (1997) suggested that environmental conditions mainly affect the growth of the microorganism present in the digestive system of adult reduviid predators. The gut microflora represents all the aspects of microbial relationships from pathogenic to obligate mutualism. Moreover, the study of the microflora associated with the insect predators may lead to the isolation of possible pathogens, which may help to design the biological control agents (Pankaj *et al.*, 2003). The present study dealing with the gut microflora associated with six different temperatures reveals that indefinite growth of microbial organisms mainly depends upon the abiotic factors, such as favoured temperatures at 25 and 30°C. We utilized the dilution plate method to recover the microorganism. This is the conventional technique used to isolate microorganisms in most of the microbiology-related studies (Santo *et al.*, 1998). Small bacterial population comprising of *Shigella* spp. like *L. casei*, and *L. dellbruckii* were recovered from the alimentary canal of the two reduviid predators. Other bacterial species from the family of *Streptococcus*, *Bacillus* and *Micrococcus* were also identified from these reduviids. From this genus common species like *Bacillus cercus, Bacillus subtilus, Micrococcus variance, Micrococcus leutus, and Enterobacter aerogenes* were isolated from the reduviids. Many members of *Enterobacteriaceae, Micrococaceea*, and *Bacillacea* are common in freshwater, soil, sewage, plants, vegetables and animals including insect guts (Mckillip *et al.*, 1997; Garcia *et al.*, 2000).

From the results, it is initiated that the bacterium is a mutualistic gut symbiont of *R. marginatus* which is vertically transmitted through the egg capsule or food and is essential for normal development and growth of the host insect. Previously, the morphology of symbiotic bacteria of plant sucking insect species has been described (Tustomu *et al.*, 2006). Although several bacteria have been isolated from the gut of some heteropterans, the nature of symbiotic bacteria has been poorly understood. Meanwhile, until today neither work nor hypothesis have been reported to this research. Among the haemolymph sucking groups of the Heteroptera, the Reduviids showed the most remarkable behavioural and anatomical arrangements for transmission of the symbiont (Gopalakrishnan, 2001).

E. aerogenes is dominantly present in the gut of *Chrysoperla rufilabris* (Burmeistre) (Mecoptera), an important biological control agent throughout the world (Harklein, 2003). In the predator *R. marginatus, Enterobacter aerogenes* is a common bacterium reported at 10°C in plants, vegetables and animals including insect guts. Temperatures alter the clima or mature bacterial community within the gut. Both lower (10 and 15°C) and higher temperatures (20, 25, 30°C and room temperqture) were in favour of the colonization of *S. aureus* and *M. variance* in both reduviids. These may be due to interspecific competition among autochthonous gut bacteria of the reduviids. All these bacteria are involved in the digestive process. Nutritional symbiosis with microbes is documented in predatory beetles and other entomophagous insects (Woolfolk et al., 2004; Lundgren et al., 2007; Lehman et al., 2009; Lundgren, 2009), but has not been well explored for Reduviidae. A more comprehensive examination may reveal important mutualistic interactions between the microbial community initially noted here and *R. marginatus* and *R. fuscipes*. It may even be that the physiological conditions of the gut that foster the enteric microbial community contributed to the slow digestion of ingested *C. cephalonica*.

The consequences of temperature-mediated symbiont loss for host organisms are not well documented. Our study has two important limitations, shared with most research on the consequences of climate for micro- and macroorganisms. First, we used constant temperatures, which do not occur in natural conditions. Second, experiments such as ours cannot evaluate the extent to which hosts or symbionts may evolve and adapt to new conditions or develop novel associations with other symbionts or hosts. Nonetheless, this work in conjunction with studies from other host-symbiont interactions suggests that climate may be an important mediator of the ecology and geographic range of many insect groups through their symbiotic relationships with microbes.

Protease activity was high at 25°C in fore- and hindgut of both predators, whereas amylase and invertase activity was highest at 35°C. In *R. marginatus,* foregut and hindgut showed maximum esterase activity at 20°C. But in *R. fuscipes*, esterase activity was greatest and equal both at 30°C and at room temperatures. THMB increased upto 30°C in reduviids. From the study result it is very clear that predators maintained at room temperature had maximum gut weight with high levels of bacterial density. Between the two predators *R. fuscipes* had minimum bacterial density compared to *R. marginatus*. In low temperatures (10, 15 and 20°C) the microbial colony-forming tendency was low in both the reduviids. Bacterial species like *B. cereus, S. aureus*, and *M. variance* increased when temperature was raised. Though both the predators have similar kinds of bacterial populations such as *K. pneumoniae, L. delbruckii* and *L. casei* they were considered as autochthonous bacteria of these reduviids and their populations present only in *R. fuscipes* implies the species specificity of bacterial populations. *S. aureus* was present dominantly between 10-20°C whereas *Micrococcus variance* was high at < 30°C. More than 12 bacterial species were identified in *R. marginatus* and *R. fuscipes* when fed with *S. litura*. Isolated bacterial strains were able to produce cellulase and amylase followed by xylanase. Maximum hydrolytic activity was observed in cellulase and amylase producing isolates belonging to *M. variance* and *S. aureus*. A more comprehensive examination may reveal important mutualistic interactions between the microbial community initially noted here and predators. It may even be that the physiological conditions of the gut that foster the enteric microbial community contributed to the slow digestion of ingested preys.

Chapter 6

ANTIGEN AND ANTIBODY INTERACTIONS

ABSTRACT

The simplicity of screening protocols associated with antigen-antibody based assays has been allowed large-scale field analyses of predator-prey interactions. This study showed haemolymph and gut possessed immune responses of the two predators. It was shown that the antigenic protein replied immuno-reactive absorbant value was high in *R. marginatus* on *S. litura* followed by *C. cephalonica* and *D. cingulatus* at 30°C. Similarly, in *R. fuscipes,* elevated response was also noted at 30°C for all the three pests (0.75 ± 0.01, 0 .68 ± 0.02, 0.54 ± 0.02 for *S. litura, C. cephalonica* and *D. cingulatus.* Results revealed that protein content increased gradually up to 20°C, and then declined at 25 and 30°C, while *R. marginauts* was fed with *C. cephalonica* and *S. litura.* However, in *R. marginatus* fed with *C. cingulatus,* adults revealed that immune response increased linearly from 10 to 30°C and then the response declined at 35°C. As in the case of *R. fuscipes* fed with *S. litura,* gut immune response did not have more variation between 10 and 20°C, even though it suddenly increased up to 20 to 30°C, again slowly declining towards at 35°C. This kind of similar immune response was observed when fed with the other two pests. It is also concluded that *R. fuscipes* haemolymph was more immunoresponsive than *R. marginatus.* The data presented here suggest that there is a huge discrepancy in the sensitivity of gut content and haemolymph-possessed immune response developed in two reduviid species.

6.1. INTRODUCTION

Serological techniques are based on antibody-antigen reactions between a specific antiserum generated against the species of interest, and antigens within potential enemies, in particular their digestive tracts, that may have ingested the species of interest. Use of ELISA (indirect Enzyme Linked Immunosorbent Assay) in nutritional studies was suggested by Hagler and Naranjo (2004) and enables us to rapidly screen the predators to obtain accurate data on gut content (Van Weeman and Schuurs, 1971; Sundarland, 1988; Greenstone and Trowell, 1994). The most important factors considered in the development of an antibody-based assay are the levels of sensitivity and specificity achieved (Sheppard and Harwood, 2005). Laboratory optimization is necessary to quantify the rates of antigen decay, the effects of temperature on decay rates, the consequence of alternative prey consumption on detection periods, and difference in detection limits between the predators. All these factors can influence the detection of prey material (Sunderland, 1996; Santos-Neto et al., 2010). Predator identification should be preferentially performed using parts of the predators' digestive tract. Tests with the gut showed a low ELISA reading, probably because the food had already suffered enzymatic action in this region. Once the ingested insect preys are exposed to the proteases in the digestive tract of the predator, antibodies in the antiserum are unable to recognize the antigenic proteins from the preys. The crop, which corresponds to a storage organ of the gut of the predator, gave the maximum ELISA reading (0.61 to 0.72 for *R. marginatus* and .0.68 to 0.73 for *R. fuscipes*).

Furthermore, the mass collection of arthropods' forgut content analysis can yield false-positive data due to surface level contamination with target prey or increased interaction between the predators and prey due to inappropriate sampling protocols (Harwood and Obryckii, 2005a, b). The simplicity of screening protocols associated with antigen-antibody based assays has allowed large-scale field analyses of predator-prey interactions (Harwood et al., 2004). This technique of using pest specific antibodies was pioneered in early 1998 (Hagler at al., 1992; Hagler and Naranjo, 1999 a, b). In the largest gut-content study reported (Harwood et al., 2001) assayed predators by indirect ELISA. Most investigators employing gut content immunoassay have used whole body homogenates for their assays (Fitcher and Stephen, 1981; Hagler et al., 1992; Hagler and Naranjo, 1994a, b; 1997). Microscopic gut content analyses are easy and affordable, but they are ineffective for most predators because the prey was liquefied (or) chewed into tiny unrecognizable pieces (Miles, 1972).

Visual identification of gut content revealed the feeding preferences of coleopteran predators (Forbes, 1983; Sopp et al., 1992) of coccinellid feeding on pollens or aphids. Furthermore, the additional investigation had developed an ELISA, which employs a species- and antibody-specific way for examining predators of the *Pectinophora gossypiella* (Saunders) on the pinkbol worm eggs (Hagler *et al.,* 1994; Hagler and Naranjo, 1997). Valuable information can be gathered and gut dissection has enabled the identification of prey remains from museum specimens. Though reduviids are good biological control agents (Ambrose, 1999; Sahayaraj, 2007), no information is available about the usage of gut content analyses using ELISA. This study aimed to record the impact of temperature and prey on the gut content analysis of two biological control reduviid predators using ELISA.

6.2. METHODLOGY FOLLOWED

Twenty to twenty five adults of *R. marginatus* and *R. fuscipes* (> 15 days old) were maintained separately in environmental chambers at 10, 15, 20, 25, 30 and 35°C on *C. cephalonica, S. litura* and *D. cingulatus* separately. Predators were removed from the environmental chambers after 15 days and immediately frozen at -20°C. Each predator's alimentary canal was separated and was homogenized in 500 µl of PBS (pH 6.8) and assayed for *S. litura, C. cephalonica* and *D. cingulatus* remaining in the gut and haemolymph of the predators. Antibody was determined through immunoassays performed into round-bottom wells of polystyrene microtitre plates coated with 500ng of *S. litura, C. cephalonica* and *D. cingulatus* purified protein separately. After antigen sensitization, the free reactive sites of the wells were blocked with 1% BSA.

6.2.1. Haemolymph collection and preparation of antigen

Haemolymph was collected from 6 to 10 adult predators by cutting the base of the sacpe region with sterilized; care was taken to avoid mixing of haemolymph with body fluid. Saturated phenylthiourea (2 µL) was added to the pooled haemolymph sample to prevent coagulation. Samples were centrifuged at 12000 g for 3 hours and the supernatant used for ELISA analysis.

In a small beaker, 25 mL of distilled water was taken and placed on a dialysis membrane for 10 to 20 minutes and then replaced in to boiled water for 10 minutes in order to remove impurities, unwanted proteins and any inactive enzymes. Care was taken to see the dialysis membrane did not touch the wall of the beaker. The leakage of the membrane was checked using a squeeze bottle. Desired length was selected and both inner and outer sides were rinsed with distilled water. One end of the membrane was fastened securely with the thread. Whole body of three pests (*S. litura, C. cephalonica* and *D. cingulatus*) were homogenized with 500 µL to 1 mL of cold PBS solution then the crude content was passed through the dialysis membrane (29 mm) and placed inside a beaker containing 500 mL of PBS. Open end of the membrane was closed securely by cotton thread. The membrane was then placed in 500 ml of PBS buffer solution and dialyzed overnight at 4°C. During this process PBS buffer was changed 3 to 4 hours in order to remove the impurities. Finally the purified protein adhering to the inner surface of the membrane was used as pest antigen.

6.2.2. Indirect Gut content ELISA

Alimentary canal samples were prepared for ELISA by homogenizing individual predators using 500 µL phosphate buffered saline in a 96-well assay plate, which were coated separately with a 100µL of aliquot for each antigen sample and incubated at 4°C overnight. The unbound antigen was discarded from the assay plate and 300 µL of 1.0% (10.0 mg/ µL) BSA in distilled H_2O was added. It was allowed to sit for 30 minutes at room temperature to block any unoccupied protein binding sites in the wells. Wells were rinsed three times with PBS–Tween 20 (0.05%) and twice with PBS. Fifty micro liters of the pest antigen was then added separately to each well (Hagler *et al.*, 1994). Then the plates were incubated for 1 hour at room temperature, then rinsed in the same manner described above. Aliquots (50 µL) of anti-rabbit's IgG conjugated to alkaline phosphates diluted 1:500 in 1.05% BSA was added to each well of the plates and incubated for 1 hour. Plate contents were discarded and again plates were rinsed three times as described above. A 50 µL aliquot of substrate solution was added to each well using the regents supplied in an alkaline phosphate substrate kit (Nune, Mexisorp, UK) following the addition of 50µL of 2N H_2SO_4. Then absorbance of each well was measured using a SLR 36 ELISA strip reader (Glaxo, Mumbai) at 450nm. Each pest's antigen was considered as a positive sample separately.

6.2.3. Predator total protein content in Indirect ELISA sensitivity

Predators devoid of the pest's antigen were frozen at -40°C for 3 days. Separately stock solutions were prepared by homogenizing 10 to 15 (*D. cingulatus*), 9 to 17 (*C. cephalonica*) and 3 to 10 (*S. litura*) with to 5 to 10 ml of PBS. 50 µl aliquot of this stock solution was equivalent to a single pest's antigen. 50 µl aliquot of stock pest antigen solution was added to each predator sample for total volume of 500 µl. Six to ten predators were homogenized in 500 µl PBS and treated as a negative control. The total protein content of each individual was determined by standard method (Bragdon, 1976). Then the samples were assayed for indirect ELISA as described. The mean ELISA absorbance value was recorded for each temperature treatment separately using ELISA reader (Glaxo, India).

6.3. OBSERVATION

Protein content of *Rhynocoris marginatus* in relation to three pests such as *S. litura, D. cingulatus* and *C. cephalonica* were significantly different (P < 0.05). For instance the overall mean protein concentration was 400, 330, 400, 475 and 425 µg/insect for *R. marginatus, R. fuscipes, S. litura, D. cingulatus* and *C. cephalonica* respectively. No negative control was maintained. The standardized ELISA consisted of homogenate each predator, regardless of its total protein content, in 500 µl of PBS with *S. litura, D. cingulatus* and *C. cephalonica* as pest antigen. Pest antigen was detected in every *R. marginatus* sample that was spiked with a single pest yielding a mean ELISA absorbance value. The *R. marginatus* samples values increased with pest antigen were immuno reactive. Here, also a standardized indirect gut and haemolymph content ELISA was used for all temperature reared predators, because we would like to find out the qualitative feeding behavior of these predators in relation to various pests.

The first step was standardized by coating the pest antigen predator/500 µl of PBS in the indirect ELISA plates. However, it reveals that irrespective of the prey consumption, a 500-µl dilution yielded maximum protein content both in *R. marginatus* and *R. fuscipes*. To minimize the high frequency of the ELISA false-negative reactions, an equivalent amount of pest antigen was added to *R. marginatus* and *R. fuscipes* samples that were homogenized in 500 µl (500 µg/well) to 1500 µl PBS, then each sample was analyzed by an indirect ELISA. From the observations we understood that a single well, which contained 100 µl homogenate was required for this study (Table 12).

Table 12. Effect of constant temperatures (^{0}C) on ELISA versus qualitative haemolymph analysis (Absorbance OD/450nm) *R. marginatus* and *R. fuscipes* with preys

Pests	Constant Temperatures (°C)						
	10	15	20	25	30	35	RT
R. marginatus							
C. cephalonica	0.43	0.45	0.50	0.66	0.63	0.43	0.61
D. cingulatus	0.41	0.44	0.50	0.63	0.62	0.42	0.62
S. litura	0.46	0.52	0.52	0.69	0.72	0.49	0.72
R. fuscipes							
C. cephalonica	0.42	0.42	0.44	0.68	0.68	0.56	0.68
D. cingulatus	0.40	0.41	0.42	0.57	0.54	0.48	0.55
S. litura	0.44	0.55	0.59	0.73	0.75	0.55	0.75

6.3.1. Effect of predator protein content on ELISA sensitivity

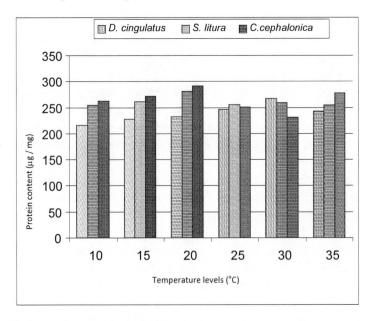

Figure 9. Fate of mean protein content of *R. marginatus* fed with *S. litura, C. cephalonica* and *D. cingulatus*

Mean protein content of *R. marginatus* fed with three pests is illustrated in Figure 9. Results revealed that protein content increased gradually up to 20°C, and then declined at 25 and 30°C when *R. marginauts* was fed with *C. cephalonica* and *S. litura*. However, protein content was gradually increased up to 30°C in *R. marginatus* and *R. fuscipes* fed with *D. cingulatus*. Similar trend was also observed in *R. fuscipes* fed with *S. litura, D. cingulatus* and *C. cephalonica* (Figure 10). The indirect ELISA was unreliable in detecting an immune response for two heteropteran predators.

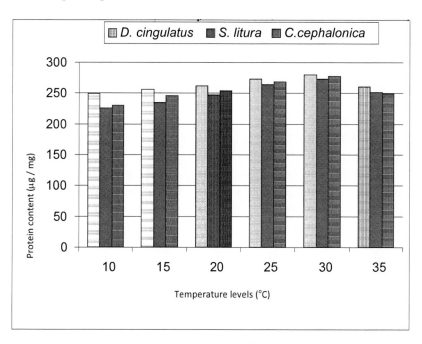

Figure 10. Fate of mean protein content of *R. fuscipes* fed with *S. litura, C. cephalonica* and *D. cingulatus*

6.3.2. ELISA response on gut of the predators

R. marginatus was maintained at a lower temperature threshold (≤ 25°C) after *S. litura* feeding and showed more of an immune response than the individual reared at 35°C. There was an irregular manner of positive response recorded at all temperatures on the tested pests, but a declined level was visibly found between 25 and 35°C. In *D. cingulatus* fed individuals more of an immune response was recorded between 20 and 30°C, and then it declined at 35°C. However, in *R. marginatus* fed with *C. cingulatus* adults revealed that immune response increased linearly from 10 to 30°C and then the response declined at 35°C. As in the case of *R. fuscipes* fed with *S. litura,* gut immune response did not have more variation between 10 to 20°C. Even though it suddenly increased at 20 to 30°C, it again slowly declined towards 35°C; this kind of similar immune response was observed when fed with the other two pests. Among the three pests, *S. litura* provided more gut immune response from these two reduviid predators than *D. cingulatus* and *C. cephalonica.*

6.3.3. ELISA response on Haemolymph

In another study, immune response was recorded using haemolymph of the predators. It was shown that the antigenic protein replied immuno-reactive absorbant value was high in *R. marginatus* (0.72 ± 0.02) on *S. litura* followed by *C. cephalonica* (0.63 ± 0.03) and *D. cingulatus* (0.62 ± 0.03) at 30°C. Similarly in *R. fuscipes,* high response was also noted at 30°C for all the three pests (0.75 ± 0.01, 0 .68 ± 0.02, 0.54 ± 0.02 for *S. litura, C. cephalonica* and *D. cingulatus* respectively). In all other temperatures revealed the immune response was more or less equal for all the pests of these two predator species. Among the three pests, maximum response was observed on *S. litura* followed by *C. cephalonica* and *D. cingulatus* (Table 12).

6.4. DISCUSSION

Various parameters, such as pH, salt concentration, temperature, time, and volume influence the precipitin reaction in most antibody-antigen systems. The results of this study suggest that selective prey consumption of the reduviid predators with optimum temperature reflect that these reduviids preferred *S. litura.* Results also confirmed that haemolymph possesses more immune response, which immune reactive property (or) tendency normally decay (or) disrupt at higher (< 35°C) and lower temperatures (> 15°C). Results also revealed that the immune response of both predators fluctuated according to the temperature regimes quoted by Hagler and Naranjo (1997). Other findings of indirect ELISA revealed that the predator gut immune response was decreased and increased based upon meal size, suitable prey consumed, temperature regimes and prey detective interval (McIver, 1981; Sunderland et al., 1987; Sopp and Sunderland, 1989). Most of the studies attributed interspecies differences in the prey detection to variable metabolic rate as a function of time and temperature (Engval and Perlman, 1971 a, b; Fitcher and Stephen, 1981; Hagler and Cohen, 1990; Sopp et al., 1992; Greenstone and Hunt, 1993). From this result it is clear that ELISA study can also be considered to know the temperature dependant immune reaction of both *R. marginatus* and *R. fuscipes*. It is also noticed that there was a rapid decline with preys like *S. litura* and *D. cingulatus*. In addition, Hagler et al. (1994) recorded considerable species gut content immune response variation present on predators' gut by immunoassays performance. Previous studies of Sahayaraj (2000), Sahayaraj et al. (2004) showed that when *R. marginatus* was provided with *H. armigera* and *S. litura,* the reduviid preferred *S. litura*. Prey preference studies also revealed that *R. marginatus* preferred mostly *S. litura* to *D. cingulatus* (Sahayaraj, 1994).

From the present study it is noted that both the reduviid predators' possessed haemolymph and gut content by ELISA detected immune response was adversely affected and depends upon the lower as well as higher temperatures. These results are also in conformation with the report of Sopp and Sunderland (1989) and Carbs et al. (2007). This indirect ELISA study clearly indicated that between the two predators, (145 mg) (0.072µg) gut content of reduviid *R. fuscipes* revealed more immune response than the larger predator *R. marginatus* (310 mg) (0.068µg). A similar trend was recorded by Hagler and Naranjo (2004). Stuart and Greenstone (1990) reported that optimal concentrations of reagent were determined through sequential check board of primary antibodies followed by primary antibody versus standard antigen dilutions; this leads to standard curves generated in an indirect ELISA. The results reveal that the marked differences of pest specific with higher immune response explained mainly depends upon the given temperature regimes. For instance the response was in favour at 25°C for *S. litura*, 30°C for *C. cephalonica* and at 35°C for *D. cingulatus,* which concordant observation was made by Ma *et al.* (1984). The prey consumed by small predators containing less protein have a greater chance for attaching or adhering to the ELISA microplate matrix than the prey consumed by large protein - rich predators. The total protein concentration present in the ELISA samples should not exceed 125 µg / samples to minimize the probability of ELISA false negative reaction. This relationship suggests that there is a rapid initial decay of ELISA sensitivity that occurs as protein content increases (Sundarland *et al.,* 1978; Hagler and Naranjo, 1994a, 1994b). In effect, the extraneous, non-target proteins associated with large predators "block" the targeted prey proteins from binding on to the ELISA matrix (Pickel, 1981). The net result is a higher frequency of false-negative reactions with large predators (Ma *et al.,* 1984; Shepard et al., 2004).

To sum up, factors such as variable predator digestive rates (Symondson and Liddel, 1993), prey sizes (Sopp and Sunderland, 1989), temperature (Hagler and Naranjo 1999a, b), predator metabolic status and developmental stage of the prey

(Hagler *et al.,* 1992) can all effect the quantitative outcome of immunoassays (Sunderland, 1996). However, very few investigations have considered the total protein content present in the samples as an important variable which affect the qualitative and quantitative outcome of indirect immunoassays.

Reduviids were subjected to various constant temperatures and fed with *C. cephalonica, S. litura* and *D. cingulatus.* Of the three pests used, *S. litura* fed individuals were more immuno-reactive at both 25 (0.61) and 30°C (0.63) in *R. marginatus* and *R. fuscipes* (0.69 and 0.71 at for 25 and 30°C) than *C. cephalonica* and *D. cingulatus.* The results showed that remaining temperature had an immune response positive but lower than that of 25°C, 35°C. It is also concluded that *R. fuscipes'* haemolymph was more immunoresponse than that of *R. marginatus.*

Chapter 7

SDS-PHAGE AND PCR TECHNOLOGY

ABSTRACT

Genetic diversity within two reduviid predators such as *R. marginatus* and *R. fuscipes* and their relationship with six different temperatures has not been extensively examined with molecular markers. Insects comprise the largest species composition in the entire animal kingdom and possess a vast undiscovered genetic diversity and gene pool that can be better explored using molecular marker techniques. RAPD analysis showed that the two reduviid predators *R. marginatus'* and *R. fuscipes'* whole body contents were clustered differently besides the Abiotic stress of temperature. Cluster analysis revealed groupings among accessions; however, no apparent correlations with putative with biotic stressed especially temperature treated predators origin were detected. Genetic variation within and between six different temperatures maintained for two different reduviid predators such as *R. marginatus* and *R. fuscipes* were investigated using RAPD (Random Amplified Polymorphic DNA) primers. The result also revealed that the examined dendrogram-based cluster analysis, which was generated from both predators with 3 various primers, KTG- 3, KTG-5 and OPE-8 denoted that the tree-constructed using the KTG-5, KTG-5 were shown variable than OPE-8 and in *R. marginatus.* In another way in the predator *R. fuscipes*, KTG - 5 and OPE 8 showed identical variation but there was no remarkable variation between 10°C and 35°C. In RAPD banding profiles of different temperatures (10-35°C) differ from each other in terms of both the numbers as well as the size of the amplified fragments with size and MW ranging between 2500 and 100bp, and between 1500 bp and 4300 bp in *R. marginatus* and *R. fuscipes* respectively. As a result, dendrogram showed GS index value was 0.02% to elevated percentage of deviation 0.54% observed away from initial GS index value at RT (0.88) KTG-5, 0.28% away from initial GS (0.95) with RT-10°C in KTG-3 primer followed by more deviations such as 0.57% genetic similarity observed between RT to 10°C.

At the same time an insufficient variation of 0.17% was observed among these six temperature regimes in OPE-8 primer. Results of this study suggest that the level of genetic diversity between these two selected beneficial predatory insect accessions should be sufficient for developing an intraspecific mapping population.

7.1. INTRODUCTION

The DNA marker technology finds useful applications of these markers especially in molecular ecology research in insects (Briggitte and Simon, 1998; Williams et al., 1990; Subodh et al., 2010). Detection and quantification of predation patterns is complex, yet essential to the development and improvement of conservation biological control (Symondson et al., 2002). Over the last 15years or so, DNA markers have made a significant contribution to the rapid rise of molecular studies of genetic relatedness, phylogeny and population dynamics based upon the abiotic stress, gene and genome mapping in insects (Hoy, 1994; Heckel, 2003). Biochemical markers that are unique to the prey species, either proteins or nucleic acids, offer a versatile means for predation detection and quantification. Previously, Hoogendoorn and Heimpel (2001) highlighted the recent trends of applications of molecular markers in insect studies and explored the technological advancements in molecular marker tools and modern high genotyping methodologies throughout that may be applied in entomological researches for better understanding of insect ecology at the molecular level. Earlier molecular profiling provides a rapid means of quantifying prey diversity within predators but when there are specific prey DNA targets with group specific primers is the principal method of choice (Symondson, 2002). This is fine for sample laboratory studies, but when there are multiple potential target prey species (Sheppard et al., 2004) and fragments (Hoogendoorn and Heimpel, 2001) the time required to assay each predator potential target becomes limited.

In field studies, the mean number of prey items in a generalist predator gut may be as a few separate PCR assays evaluated (Harper et al., 2005). This technique is effectively peculiar for many useful field-based ecological studies. Rapid PCR – based screening systems needed for the study of the prey diversity of generalist predators have been developed to expend the potential of molecular detection in to various areas of food web research (Dodd, 2004). From the published results, it becomes clear that except for the haematophagous reduviids such as *Trypanosoma rangeli* and *Trypanosoma cruz* (Vallejo et al. 1999), to date, no information is available on the polyphagous reduviid predators. The PCR technique was developed by Ehrlich in 1989. It is one of the simplest, fastest and least expensive molecular approaches, which makes use of RAPD–PCR (Randomly Amplified polymorphic DNA) (Shappiro et al. 1988) to amplify a region of DNA that lies between two regions of known sequence (Teresa et al. 2002). PCR method has been used in many fields including understanding the genetic variability in insects (review of Sheppard and Hardwood, 2005). The isolation of DNA from insects normally does not present any specific problems. Therefore, any one of a multitude of techniques used for isolation of DNA from other organisms will usually work with insect tissue (Reza et al., 2008).

Therefore, it is envisaged to analyze quantitatively and electrophoretically in relation to temperature modification also. This chapter deals with changes of whole body and egg macromolecules (carbohydrate, protein and lipids) of two reduviid predator species by spectrophotometry whole gut protein by electrophoresis method; gut DNA polymorphism by RAPD–PCR analysis by AGE in relation to six different constant temperatures on *R. marginatus* and *R. fuscipes*.

7.2. METHODOLOGY FOLLOWED

7.2.1. Protein sample preparation for Electrophoresis

Six to ten adults of *R. marginatus* and *R. fuscipes* were taken separately from the stock insects, which were maintained in different temperatures including the room temperature categories. Foregut was dissected out from the predators and homogenized in a homogenizer. Eppendrof tubes containing 75µl of gut sample was boiled at 50-60°C for 3 minutes and allowed to cool at room temperature. The sample was then centrifuged at 10,000 rpm; supernatant was collected and used as the protein sample for electrophoresis.

7.2.2. Procedure for Electrophoresis

SDS polyacrylamide slab gel electrophoresis was carried out using the method of Laemli (1989) with minor modifications. A sandwich was made with two glass plates separated by spacer strips. The spacer strips were coated with Vaseline for adhering mechanisms. The glass plate was kept vertically by placing it on a stand, which could hold the plates vertically. A few ml of distilled water was poured between the plates to check for leakage if any. The resolving gel of 12% (P^H7.6) was poured in to the space between the glass plates after removing distilled water. The level should be about 2cm below from the notch. It was kept for polymerization for about 30 minutes then made a layer of distilled water on the surface of the resolving gel, to avoid the contact between the gel surfaces, air and also to make an even surface.

When polymerisation was completed the distilled water and stacking gel were poured (7.5% P^H7.6) over the resolving gel and the Teflon comb with fingers (each finger with 7cm wide) was inserted into the gels, and allowed to polymerize for 30 minutes. After polymerisation, the glass plates were clipped out from the stand and also, the bottom Teflon spacers were removed. The slab gel was made clean with filter paper and the plate was attached to the electrophoresis apparatus. The electrode buffer (TBE) was poured to the lower and upper chamber. Then the Teflon comb was carefully removed from the gel, supernatent of the previously prepared sample was added in to each well at a volume of 15µl with the help of a microlitre syringe. Marker protein of 14-100 KDa (Genei, Bangalore, India) was loaded in one well as a reference. Initially a current of 60V was supplied with the sample entered into the separating gel and electrophoresis was continued at 120V till the marker dye reached the bottom of the separating gel (resolved gel). At the end of electrophoresis, run glass plates were gently moved apart with a spatula, by running a stream of electrode, the gel in to a solvent resistant plastic trough for staining (Coomassive brilliant blue- CBB) for overnight and destaining (24 hrs) until clear band could be seen.

7.3. DNA EXTRACTION AND AMPLIFICATION

Six to ten reduviid adults of the predators were reared at different temperature regimes for more than a month randomly selected and homogenized with 0.5 ml extraction buffer (8% DTAB, 1.5M NaCl, 100mM Tris, 50M EDTA 10% SDS and proteinase K) and ground further. The extract was incubated for 2-3 hrs at 50^0C–60^0C to allow the separation of DNA also for the denaturation of proteins. The mixture was centrifuged for 5 minutes at 10,000 rpm. The supernatant was cleaned away from the protein and lipids by phase extraction with an equal volume of phenol, chloroform and iso-amyl alcohol (25:24:1). DNA was precipitated by adding one-tenth volume of 3M NaCl + two volume ice-cold 95% absolute ethanol and incubated for one hr at -20°C. The precipitated DNA was centrifuged and then washed with 70% ethanol; DNA was vacuum-dried and resuspended in 100µl TE buffer, pH 8.0). The concentration and purity of extracted DNA was determined spectrophotomectrically (UV- instrument) at 260 nm and 280 nm. Samples showing the one OD (optical density) equivalent to 50 µg and purity (determined by the ratio of 260 nm and 280 nm) 1.5 to 1.8 alone were taken for further analysis. Template DNA extracts were stored at -20°C and thawed at room temperature for further amplification.

7.3.1. PCR Amplification

The extracted DNA from the experimental predators was subjected to PCR analysis using six universal primers. Among the six, three primers namely KTG-3-(5'-GTAGACCCGT-3'), KTG-5 (5'-AACGCGCAAC-3') and OPE 8-(5'-AACGGCGACA-3') (GENEI scientific supplies, Bangalore) were selected for proceeding towards amplification. PCR reactions were performed in 25 µl of reaction mixtures containing 1 mM dNTP mix (5.0 µl), 1.0 µl template DNA (50 ng / µl), 10 mM RAPD primer (2.5 µl), 10X Reaction buffer, 25 mM Mg Cl$_2$ (1.5 µl), 2.5 units of Taq polymerase enzyme (5U / µl) (Bangalore Genei, India) and sterile de-ionsed water. The 25 µl of reaction mixtures referred to above were placed in PCR tubes in two layers. The bottom layer consisted of all reagents except *Thermus aquaticus* (Taq), sterile distilled water and sample DNA. Amplification was performed with thermocycler (Master cycler ep's Eppendrof, India) for 40 cycles. Thermal cycles were programmed for initial denaturation at 94°C for 2 minutes. Each cycle consisted of 40 seconds annealing at 94°C followed by 72°C for 5 minutes for final extension. Amplified, samples were stored at 4°C prior to electrophoresis. PCR amplified products were separated on 1.4% agarose gels submerged in 1X TBE, and the banding profiles were visualized with ethidium bromide. Gels were documented using Biotech documentation and analysed with Gel Del TM software (Bangalore) (Carezza Booto *et al.,* (2005). Genetic similarity and dissimilarity dendrogram was made from the similarity data using UPGMA method of the programme Digital Gel Documentation (Biotech, Tamil Nadu, India).

7.3.2. Analysis of data

Using three selected primers with six randomly selected temperatures both treated predators were subjected to comparative analysis. RAPD patterns and gut protein polypeptide profiles were visually analyzed and scored bands were noted through the photographs. For the analysis and comparison of the patterns a set of distinct, well-separated bands were selected. The genotypes were determined by recording the presence (i) or absence (o) of these bands and neglecting other (weak and unsolved groups) bands. Genetic identity (GS) and genetic distance (DS) values between the total six temperatures were calculated using the data generated from the RAPD profiles using digital gel documentation (Biotech, Tamilnadu, India). Genetic distance values were utilized to contract to dendrogram through clustering analysis of Unpaired Group Method of Arithmetic Averages (UPGMA) to determine the relationship among the six various temperatures on the predator of *R. marginatus* and *R. fuscipes.*

7.3. RESULTS

7.3.1. Temperature and gut protein profile of reduviids

Rhynocoris marginatus adults gut protein analysis by SDS PAGE showed that both 35, 30°C produced 5 polypeptides, whereas 25, 20°C and 15°C produced seven bands with molecular weight between 6.5 to 205 kDa. Two molecular weight polypeptides such as 56.0 and 205 kDa were absent at 10°C in *R. marginatus* (Plate 5a). From the dendrogram analysis revealed that gut protein profile showed higher genetic identity (GS) at 20°C (0.83), than the low temperature regimes such as 10, 15°C (GS = 0.77). As in the case of dissimilarity (Genetic distance-GD), minimum value was recorded (0.16) at temperatures of 35, 15 and 20°C. It was further reduced when the reduviid was subjected between 25 and 30°C (0.11).

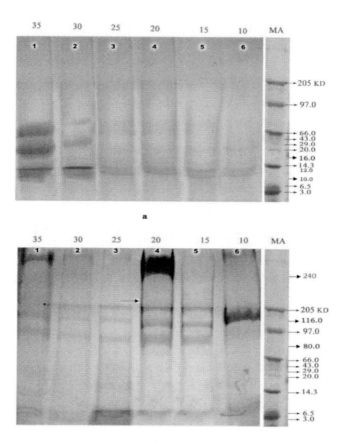

Plate 1. Whole gut SDS-PAGE protein profile of R. fuscipes and R. marginatus in relation to constant temperatures - 10, 15, 20, 25, 30 and 35°C

Plate 1 depicts gut protein profile of *R. fuscipes*. Four uniform polypeptides appeared at 35, 30, 25 and 20°C (between the molecular weight range 2.0 to 205 kDa) except at 10°C. However, at 35 and 20°C one polypeptide (<205 kDa) was peculiar, in gut protein owing to absence of the higher molecular weight polypeptide. The dendrogram analysis showed the highest genetic similarity between 25 and 30°C (0.81). Lowest genetic similarity (0.22) was noted at 10°C and 15°C. Dissimilarity (DS) did not possess striked main variations, even though they ranged between 0.23 and 0.11. A dendrogram was predicted to find out the genetic relationship in *R. marginatus* and *R. fuscipes* when subjected to various temperature regimes. From the results it is very clear that the predators reared at 30° C had close relationship with 25°C and also another category of 10 and 15°C in *R. marginatus,* whereas in *R. fuscipes,* relationship was recorded between the temperatures 10 and 15°C.

7.3.2. DNA Amplification of the reduviids

PCR amplified products having 400 and 600 bp were common irrespective of the primers in *R. marginatus.* Such a similarity was not observed when the whole body DNA of *R. fuscipes* was amplified with OPE-8, KTG-3 and KTG-4 Primers. Interestingly OPE-8, KTG-3 and KTG-5 produced a unique bp product of 1200-150 and 50 bp in *R. marginatus.* Similarly these 3 primers produced 950, 200 and 300 and 900 bp in *R. fuscipes.* The present study reveals that RAPD markers were efficient for the assessment of genetic similarity and dissimilarity coefficient using Digital Gel Documentation between the six temperatures within the same species described in the dendrogram. Apparently the resulting data present in Tables 19 and 20 were further processed for cluster analysis using the unweighed paired group of average method (UPGMA). Totally seven primers were tested, four primers (KTG- 1, 2 and 4 and OPE-8) yielded no clear or any scorable bands, but the remaining 3 primers (KTG –2, KTG – 5 and OPE – 8) were amplified and produced scorable polymorphic bands. Primers KTG –3 and KTG – 5 amplified maximum numbers of polymorphic bands ranged about 31 to 34, in *R. marginatus.* Primers such as OPE-8 and KTG –3 produced 26 bands each in *R. fuscipes.*

7.3.3. Genetic similarities (GS) in *Rhynocoris marginatus*

The KTG–3 primer predicted dendrogram consists of two clusters. Cluster–1 deserved a higher GS value and was higher (0.84) than the remaining temperatures. As in the case of another temperature held in cluster – II, this again stood for only one temperature at 20°C, and it consisted of an estimated GS value of 0.80, whereas, cluster II again divided into II-b1 and C-II b2. These sub clusters to the temperatures of 30 and 35°C also possessed similar GS value for 25 and 10°C. Of these four temperature regimes 15°C (C-II b2) consisted of an estimated GS value (0.70). This was 10% increas away from C-II b1, which consisted at 30 and 35°C. Finally C-II b2 represented temperatures both lower (10°C) and optimum (25°C); both the temperatures shared an equal GS value of 0.40. This result clearly shows that 40.4% deviation was observed between RT and lower temperatures of 10°C (Plate 2).

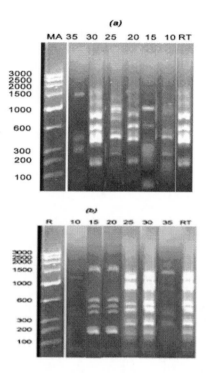

Plate 2. RAPD profile of *R. marginatus* at different temperature levels with KTG-5 (a) and KTG-3 (b) primers

While estimating the KTG-5 primers the drawn dendrogram consisted of two main sub clusters. Cluster-1 stands for moderate temperature at 30°C and possessed the GS value of 0.76. Another branch of cluster–II divided into two sub clusters such as Cluster-IIa and b. Cluster-IIa mainly influenced the two different temperatures such as 35 and 15°C. These two temperature levels shared the similar GS value 0.57 followed by the sub cluster-II b accompanied with the remaining temperature regimes such as 10, 20 and 25°C. Among these three temperature regimes, 25°C and RT indicate similar GS value that was 0.65. Similarly, the other two low temperatures at 10 and 20°C also consisted of the GS value 0.29. From the KTG-5 primers represented, initial genetic similarity index was 0.76. While the overall results clearly noted highly deviated value such as 0.53% at 35°C, 0.47% shows a low temperature 10°C (Plate 2).

When the *R. marginatus* drawn with OPE-8 primer expressed the dendrogram, it consisted of two main clusters, cluster-I and cluster-II. C-I mainly possessed the temperatures (RT) that possessed the highest GS value of 0.84. Cluster–II was broadly divided into two sub clusters namely Cluster-IIa and b.

The 35°C category included C-II b. Interestingly, both temperatures shared the same GS value of 0.75. Likewise, remaining adjacent temperatures such as 20 and 25°C stand for C-IIa and another sub cluster C-II b mainly accompanied with the temperatures of 10 and 15°C. These four temperatures (10-25°C) possessed the same value of 0.50. Here the primer predicted overall similar value was 0.34% along with the optimum deviation observed from RT. The initial GS index value was 0.84 noted at 10-25°C. This result intimates association among the temperature levels at 10-15, 20-25 and 30-35°C (Plate 3a).

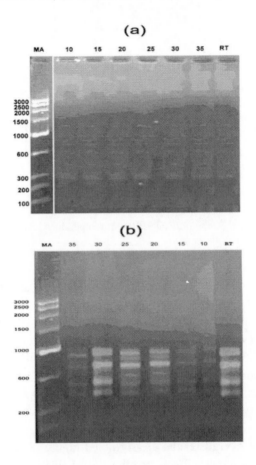

Plate 3. RAPD Profile of *R. marginatus* (a) and *R. fuscipes* (b) on six different temperatures with OPE-8 primer.

7.3.4. Genetic similarities (GS) in *Rhyhnocoris fuscipes*

The KTG-3 primer had drawn a dendrogram, which predicted results clearly for smaller cluster arrangements based upon each treated temperature level (10 to 35^{0}C). The higher genetic identity or GS was observed at higher temperature at RT (0.95) followed by 30°C (0.94), 25°C (0.93), 20°C (0.90) and 15°C (0.85). Finally standard GS value (0.67) was denoted at lower temperature (10°C) as well as higher temperature (35 °C) (Plate 4a). Despite, the six different temperature regimes reported all the genetic similarity values were similar. They showed insignificant range of deviation such as 0.01% between RT-25°C and 0.05% for 20-15°C and 0.20% at 15-35 respectively. The overall results indicated 0.28% decriational from initial GS (0.95) with RT-10°C (Table 13).

Table 13. Dendrogram analyses with three primers (KTG – 3, KTG – 5 and OPE – 8 showing genetic similarity (GS) on *R. marginatus* at various temperature regimes (°C), RT – Room temperature

Predator	Temperature (°C)	KTG – 3	KTG – 5	OPE – 8
R. marginatus	10	0.40	0.29	0.50
	15	0.70	0.57	0.50
	20	0.80	0.29	0.50
	25	0.40	0.65	0.50
	30	0.60	0.76	0.75
	35	0.60	0.57	0.75
	RT	0.84	0.66	0.84
R. fuscipes	10	0.67	0.40	0.75
	15	0.85	0.70	0.75
	20	0.90	0.80	0.87
	25	0.93	0.40	0.75
	30	0.93	0.60	0.87
	35	0.94	0.60	0.75
	RT	0.95	0.89	0.92

Apart from the primer, KTG-5 is revealed in dendrogram. This primer does not divide into clusters but they are arranged as separate accessions and stand for lower and higher temperatures (RT-10°C) instead of a cluster. In case of estimated higher GS value 0.88 observed at RT (30± 1.5°C) and the lower temperatures (10 and 15°C) been secured adjacent GS value of 0.86. In the same way, the following temperature regimes show a little bit of variation from the GS value 30°C for 0.84,

25 °C for 0.77 and 20°C for 0.66 respectively. Meanwhile, 15 and 10°C lower temperatures express a lowest GS value of 0.34. The overall results show closely related GS range was 0.02%. The elevated percentage of deviation was 0.54% observed away from initial GS index value at RT (0.88) seen among each of the six temperatures (Plate 4b).

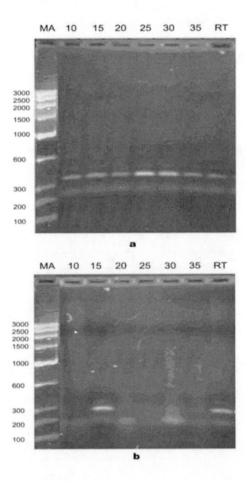

Plate 4. RAPD analysis of *R. fuscipes* with the primer of KTG-3 (a) and KTG-5 (b) at six temperature levels.

Results based on the OPE-8 primer shows cluster analysis mainly divided into two clusters: cluster- I and II. Cluster-I consists of one temperature of RT (30.1±1.5 °C) and it has also secured higher GS value of 0.92. The second cluster again is broadly divided into two sub clusters (cluster- IIa and cluster -II b). At 10 and 20°C the second highest as well as similar estimated GS value is 0.87. These two temperature levels represented Cluster-II as comprising of the remaining temperature regimes (say 35 and 15°C). The GS value was almost similar noted as 0.75 represents Cluster-IIa and C-II b (Table 13). Though, 25°C and 10°C are related to cluster- II b, corresponding temperatures also share similar estimated GS value 0.75. Observation was made in sub cluster C-IIa, too. As a result the dendrogram found more deviations such as 0.57% genetic similarity observed between RT-10°C, at the same time an insufficient variation 0.17% was revealed among these six temperature regimes (Plate 3b).

7.4. DISCUSSION

Gut protein pattern showed significant reduction in the number of polypeptides when the reduviid was subjected to difficult temperature treatments. Molecular weights of the gut protein of the two predator species ranged between 205 and 14.3 kDa and below 205 and 14.3 kDa in *R. marginatus* and *R. fuscipes,* respectively. The electrophoretic variation of the protein bands in the whole alimentary canal of the adult predators showed in Plates 6a and b. Such qualitative profile of protein observed during adult transformation with temperature confirms previous findings (Tefler *et al.,* 1983; Ryan and Dick, 2001; Teresa et al., 2002). As in the case of whole gut of *R. marginatus,* six polypeptides were evident at room temperature. Intensity of polypeptides, position, size and shape for all 6 temperatures determined slight qualitative changes between 10-15, 30-35 and 20-25°C. In *R. fuscipes* higher and lower temperatures had been peculiar with high molecular weight polypeptides. This agreed with earlier findings in Coleopteran predators that these were not based upon only temperature but endepeptidases in the alimentary canal (Agrell & Landquist, 1973; Addedire and Balogum, 1995).

A tool that is routinely used to investigate predator–prey relationships is genetic gut content analysis (GCA) of the predator using polymerase chain reaction (PCR) primers that amplify unique regions of the prey's genome (Symondson, 2002; Greenstone et al., 2007; Harwood et al., 2007; Harwood et al., 2009; Lundgren et al., 2009a; Weber and Lundgren, 2009a,b), but this technique has not heretofore been used to study the impact of temperature on any insect. Predators digest these DNA markers over time, and the rate at which this occurs is dependent on a range of environmental factors including temperature and inherent qualities of the food itself (Zhang et al., 2007; Durbin et al., 2008; Weber and Lundgren, 2009b). The complexities of this digestion process hamper interspecies comparisons of predator efficacy and predation rates under field conditions. Hence we considered to study the temperature impact on predatory through PCR analysis. The inter-population profile of the six temperature combinations showed a remarkable banding pattern of powerful band, high intensities of genetic variation or suggesting heterogenous as well as homogenous amplified DNA between 10 to 15, 20 - 25, 25 - 30 and finally 35°C encrisafing faint banding pattern which ranged between then 2000 to 200bp MW and 1000 to 300bp MW in *R. marginatus* and *R. fuscipes* respectively. In *R. fuscipes,* KTG-5 and KTG-3 primers produced amplified products that were homogenous with respect to all temperatures. This result demonstrates precisely KTG and KTG-3 primers in *R. marginatus* adults revealing that weak and powerful band between 10°C and 35°C and in OPE–8; we recorded low molecular bp bands with uniform pattern.

This result clearly indicated high and low temperatures caused changes in the DNA pattern and level both in terms of quality and quantity. The success of PCR depends on the quality of the DNA- it must be free from any contaminants and from protein, nucleus that interfere the amplification process (Greenstone et al., 2005; King et al., 2008). PCR based protocol shows great power for quick and simple characterization of genetic variation within and among the population (Whittmann et al., 2002). The PCR techniques were successfully used in studies on DNA of reduviids, *T. cruzii* (Breniere et al., 1995; Carezza Botto et al., 2005) and preliminary studies on the selection and activity cycles on heteroptera Reduviid species (Canals et al. 1997; Moser et al. 1989; Russomando et al. 1992). Qualification and quantification of DNA under such circumstances becomes necessary to amplify the DNA polymerase chain reaction and then quantify the PCR product. The amount of product would be measured as well as purity of the extracted initial DNA. Due to extreme sensitivity of the PCR reaction, however even very small variation in the reaction efficiency would result in significant differences in the amount of final product formed. RAPD analysis allowed grouping the insects in accordance with the place of capture in contrast to the molecular analysis. It is possible that this grouping could be due to the phenotypic plasticity, the expression of different phenotypes in single genotype when subjected to different environments (Whiteman, 2005).

These results indicated that high quality DNA could be isolated from both of the predators reared under a range of temperatures. In addition, successful PCR was possible when the amount of DNA specifically with temperature produced amplification product was widely varied. This demonstrates that the quantity of DNA was an initial process for amplification of isolated DNA molecules. The study had also addressed the suitability of reagents used to store the experimental adult predators for subsequent DNA isolated in deep freeze (-20°C). Storage conditions are apparently not critical for experimental samples stored less than 6 months.

Temperature mainly determined biological and biochemical functional activities including macromolecules of DNA and RNA content (Price, 1965; Dodd, 2004; Carreza Booto et al., 2005). The result also indicated that when we analyzed the dendrogram like trees diagram, which generated from both predators with 3 various primers, KTG- 3, KTG-5 and OPE-8, the tree-constructed using KTG-5 showed variation than OPE-8 and in *R. marginatus*. In another way in *R. fuscipes* predator, KTG-5 and OPE-8 showed similar variation but they did not observe more variation between 10°C to 35°C.

From this study it is understood that when the temperature reached 35°C or more or less than 10°C amplified DNA of the predators were less intense than the optimum normal temperatures banding profile. Notably, this similar trend was also observed by Garnoel and Barett (1993). Some individuals of the heat treated group of insects exhibited a faint band, indicating temperature fluctuation is also caused or accompanied with the presence of a small amount of the bacterium on the alimentary canal of an insect species reported previously by Tsuchida et al. (2002).

It is concluded that PCR is an excellent tool that can be applied to identity genetic polymorphism as well as change in the genetic constituents. It depends upon the temperature variation within and among the same individual of the predators. PCR employed here is a method, which has the large applicability of RAPD but also could generate differences in the banding pattern that are more informative for population analysis. Moreover, this result indicates prolonged temperature (highest above 35°C) stores could reduce or denature the protein molecules. In RAPD banding profiles of different temperatures (10-35°C) differ from each other in terms of the numbers as well as the size of the amplified fragments with size and MW ranging between 2500 and 100bp then 1500 bp to 4300 bp in *R. marginatus* and *R. fuscipes,* respectively.

GENERAL CONCLUSIONS AND FUTURE RECOMMENDATIONS

1. Nymphal total developmental time doubles when the reduviid is reared at low temperature (20°C) and reduced the survival rate thrice. Similarly fecundity and hatchability also reduced drastically at lower temperature. Higher temperature (35°C) abrubted the nymphal development and also disturbed the fecundity and hatchability of these reduviids. Deformities were recorded on the legs, wing, and incomplete moulting was also observed when the predators were maintained at 10, 15, 20 and 35°C. Lower and higher threshold temperatures for *R. marginatus* are 24.8 $^{\circ}$C, but it ranged from 23.92 $^{\circ}$C to 25.3 $^{\circ}$C for *R. fuscipes*.

2. The foraging behaviour of reduviid predators was greatly determined by the temperatures. However, temperature dones affects the feeding chronology of the reduviids. Predators approached their preys quickly at higher temperatures and handled them with more time. Fifth nymphal instar and adult predators were more successful in encountering the large sized preys. Though different nymphal instars of *R. marginatus* preferred life stages of lepidopteran larvae, second and third instar reduviids preferred second, third and fourth instar *D. cingulatus* and the remaining life stages of these reduviids often preferred *D. cingulatus* adult. All the nymphal instars and adults of *R. fuscipes* mainly preferred second to fourth instar larvae of *S.litura* and second to fifth instar nymphs of *D.cingulatus*.

3. Predators maintained at room temperature had maximum gut weight with high level of bacterial density. THMB increased upto 30°C in *R. marginatus* (12.37×10^{4}) and *R. fuscipes* (9.4×10^{4} CFU/gm). Between the

two predators, *R. fuscipes* had minimum bacterial density than *R. marginatus*. Bacterial species like *Bacillus cereus, Staphylococcus aureus and Micrococcus variance* increased when temperature rose. Both the predators had similar kind of bacterial populations such as *Klepsiella pneumoniae, Lactobacillus delbrueckii* and *Lactobacillus casei* and these were considered as autochthonous bacteria of these reduviids. *Staphlococcus aureus* was present dominantly between 10-20°C whereas *Micrococcus variance* was high at < 30°C.

4. Isolated bacterial strains were able to produce cellulase and amylase (C+A) followed by xylanase. Maximum hydrolytic activity was observed in cellulase and amylase producing isolates belonging to *M. variance* and *S. aureus*. Protease activity was high at 25°C in fore and hindgut of the predators whereas amylase and invertase activity was highest at 35°C. In *R. marginatus,* foregut and hindgut showed maximum esterase activity at 20°C. But in *R. fuscipes*, esterase activity was maximum and equal both in 30°C and at room temperatures. While the location was compared, these enzyme activities were well pronounced at hindgut than the foregut.

5. Of the *C. cephalonica, S. litura* and *D. cingulatus* pests, *S. litura* fed individuals were more immuno-reactive at both 25 (0.61) and 30°C (0.63) in *R. marginatus* and *R. fuscipes* (0.69 and 0.71 at for 25 and 30°C) than *C. cephalonica* and *D. cingulatus*. It is also concluded that *R. fuscipes* haemolymph was more immunoresponse than *R. marginatus*.

6. It is concluded that PCR is an excellent tool that can be applied to identity genetic polymorphism as well as change in the genetic constituents depending upon the temperature variation within and among the same individual of the predators. In RAPD banding profiles of different temperatures (10-35°C) differ from each other in terms of both the numbers as well as size of the amplified fragments with size and MW ranged between 2500 and 100bp then 1500 bp to 4300 bp in *R. marginatus* and *R. fuscipes,* respectively.

FUTURE RESEARCH AREA COULD BE CONSIDERED

1. This study can be extended to other potential biological control agents belong to Reduviidae.
2. The efficiency of natural enemies is affected by environmental conditions, mainly temperature. However, so far no information has been

available in this regard. We advise to study the effect of temperature on the functional and numerical response of life stages of reduviid predators against pests dwelling in temperate areas can be carried out in the future.

3. Constant temperature on the hormone level will be undertaken to find out the various impacts to the reduviid predators.

4. It is essential to record the optimum temperature in which various life stages of the reduviid can be stored to utilize them for agumentative biological control programme.

5. The data presented in this book on antigen-antibody interaction suggest that there is a huge discrepancy in the sensitivity of gut content and haemolymph-possessed immune response developed in two reduviid species.

ABBREVATIONS USED

^{o}C- Degree centigrade
BIPM- Biointensive integrated pest management
BSA- bovine serum albumin
CBB- Coomassive brilliant blue
DD- degree-days
DNA- De oxyribo nucleic acid
DS- dissimilarity
ELISA- Enzyme Linked Immuno Sorbent assay
GD- Genetic distance
GS- Genetic identity/similarity
HDTT- Higher developmental threshold temperatures
IPM- Integrated Pest Management
kDa- kilo Dolton
L.- Linnaeus
L: D- Light: dark hours
LDT- lower developmental threshold
LDTT- lower developmental threshold temperatures
MW- Molecular weight
NA- nutrient agar
NPV- Nucledar Polydehyudrosis Viruses
PBS- phosphate buffered saline
PCR- Polymerase Chain Reaction
RAPD- Random Amplified Polymorphic DNA
RF- *Rhynocoris fuscipes*
RH- Relative humidity

RM- *Rhynocoris marginatus*

RT- Room temperature

SDA- Serially Dilution Agar

SDS-AGE- Sodium Deudosyle Sulphate – Agarose Gel Electrophoresis

SDS-PAGE- Sodium Deudosyle Sulphate – Poly Agralamide Gel Electrophoresis

THBP- Total heteroptrophic bacterial populations

TSA- Trypticase soy agar

UPGMA- Unpaired Group Method of Arithmetic Averages

UDT- Upper developmental threshold

W/V- weight/volume

REFERENCES

Abraham, E. G., Nagaraj, J., and Datta, R.. K. 1992. Biochemical studies of amylases in the silkworm *Bombyx mori* L. comparative analysis in diapausing and non-diapausing Strains Insect. *Biochemical Molecular Biology,* 21: 303-311.

Adedir., C.O. 1984. Distribution of carbohydrares and proteases in the intestine of the Kola nut Weevil, *Sophrorhinus insphsbus* Fanrt (Coleoptera: Curculionidae) and response of proteases to inhibitors from Kola nuts. *Applied Entomological Zooogy,* 29: 331 - 338.

Adedire, C.O., and Balogun, R.A. 1995. Digestive enzyme and regional localisation of proteolytic endopeptidases in the alimentary canal of the Kola Weevil, *Sophrorhinus insperatus* Faust (Coleoptera: Curculionidae). *Entomon,* 20 (4): 183 – 189.

Agrell, I.P.S. and Landquist, A.M. 1973. Physiological and biochemical changes during insect development In (Rockstein M The physiology of insect). (Ed.) Vol. I. Academic Press. *New York, London.*

Altieri, m. a., Francis, C.A., Schoonhoven, A.V. and Doll, J.D.A. 1978. Review of insect prevalence in maize *(Zea mays* L.) and bean *(Phaseolus vulfaris* L.) polycultural systems. *Field Crops Research,* 1: 33 – 49.

Almeida, C.E, Duarte, R., Guerra, R.S., Pacheco, R., and Costa, J. 2003. *Triatoma rubrovaria* (Blanchard, 1843) (Hemiptera: Reduviidae, Triatominae). Heterotrophic resources and ecological observations of five populations collected in the state of Rio Grande do Sul. *Brazil. Memorlas do Instituto Oswaldo Cruz,* 97: 1127-1131,

Ambrose D.P. 1980 Bioecology, ecophysiology and ethology of Reduvuiidae (Heteroptera) of the scrub Jungles of Tamilnadu, Ph.D., thesis, University of Madras, Madras pp : 17 – 25.

Ambrose, D. P. 1987. Assassin bugs of Tamil Nadu and their role in biological control (Insecta: Heteroptera: Reduviidae). In: (K. J. Joseph and U. C. Abdurahiman eds.), Advances in Biological Control Research in India. Dept. of Zool., University of Calicut, Calicut, India. pp. 16 - 28.

Ambrose, D. P. 1988. Biological control of insect pests by augmenting assassin bugs (Insecta: Heteroptera: Reduviidae). In: (K. S. Anantha-subramanian, P. Venkatesan, S.Siva- raman eds.), Bicovas II. Loyola College, Madras, India pp. 25: 240-243.

Ambrose, D.P. 1995. Reduviids as predators: Their role in biological control, In: Biological control of social forests and plantation crops insects, (T.N. Ananthakrishnan (New Delhi: Oxford and IBH Publishing Co. Ltd.,) (Ed.) pp 153-170.

Ambrose, D.P. 1997. Mating behaviour of the assassin bugs *Neohaematorrhophus therassi* and *Rhynocoris fuscipes* (Hemiptera: Reduviidae). *Indian Journal of Biodiversity*, 1 (1-2): 131-139.

Ambrose D.P. 1999. Assasin Bugs, Science Ambrose .D.P. Publishers. Inc. Enfield New Hampshire Editors: USA, pp-337

Ambrose, D.P. 2001. Assassin Bugs (Heteroptera: Reduviidae) in Integrated Pest Management Programme: Success and Strategies. In: (Strategies in Integrated Pest Management, Ignacimuthu), (Eds.) S. and Alok Sen, Phoenix Publishing House Pvt. Ltd., New Delhi, pp: 73 – 85. Ambrose, D. P. 2003. Bio control potential of assassin bugs. *Journal of Experimental Zoology*, 6(1): 1 - 44.

Ambrose, D.P. 2006. A checklist of Indian Assassin Bugs (Insecta: Hemiptera: Reduviidae) with taxonomic status, distribution and diagnostic morphological characteristics. Zoos' Print Journal, 21(9): 2388-2406.

Ambrose, D. P. and Claver, M. A. 1995. Food requirement of *Rhynocoris kumarii* Ambrose and Livingstone (Heteroptera: Reduviidae). *Journal of Biological Control,* 9: 47 - 50.

Ambrose, D. P. and Claver, M. A. 1996. Impact of prey deprivation in the predatory behaviour of *Rhynocoris kumarii* Ambrose and Livingstone (Heteroptera: Reduviidae). *Journal Soil Biology and Ecology,* 16 (1):78 - 87.

Ambrose, D. P. and Claver, M. A. 1999. Suppression of cotton leafworm *Spodoptera litura*, flower beetle *Mylabris pustulata* and red cotton bug *Dysdercus cingulatus* by *Rhynocoris marginatus* (Fabr.) (Het., Reduviidae) in cotton field cages. *Journal of Applied Entomology.*123:225-229.

Ambrose, D.P. and Claver, M.A. 2001. Survey of Reduviid Predators in Seven Pigeonpea Agroecosystems in Tirunelveli, Tamil Nadu, India. *International Chickpea and Pigeonpea Newsletter*, No. 8 – 2001, 44-45.

Ambrose, D.P. and Claver, M.A. 2001. Prey preference of the predator *Rhynocoris kumarii* (Heteroptera: Reduviidae) to seven cotton insect pests. *Journal of Applied Zoological Research*, 12(2&3): 129-132.

Ambrose, D.P., and Kumaraswami, N.S. 1990. Functional response of the reduviid predator *Rhinocoris margainatus* on cotton Stainer *Dysdercus cingulatus* Fab. *Journal of Biological Control,* 4:22-24.

Ambrose, D.P. and Kumaraswami, N.S. 1993. Food requirement of *Rhinocoris fuscipes* Fab. (Heteroptera, Reduviidae). *Journal of Biological Control*, 7(2): 102-104.

Ambrose, D.P. and Maran, S.P.M., 1999a. Substrata impact on mass rearing of the reduviid predator *Rhynocoris marginatus* (Fabricius) (Insecta: Heteroptera: Reduviidae). *Pakistan Journal of Biological Sciences,* 2(4): 1088 - 1091.

Ambrose D. P, and Maran, S. P.M. 2000a. Polymorphic diversity in salivary and haemolymph proteins and digestive physiology of assassin bug *Rhynocoris marginatus* (Fab.) (Heteroptera: Reduviidae). *Journal of Applied Entomology,* 124: 315-317.

Ambrose, D.P. and Maran, S.P.M. 2000b. Haemogram and Haemolymph protein of male and female mated and oviposited *Rhynocoris fuscipes* (Fabricius) (Heteroptera: Rduviidae : Harpactorinae). *Advances in Biosciences*, 19(II): 39 - 46.

Ambrose D.P., and Mayamuthu, T., 1994. Impact of sex starvation, antennectamy, eye blingidng and tibial comb coating on the predatoru behaviour of *Rhynocoris fuscipes* (Fab.) Insecta: Heteroptera: Reduciidae), *Journal of advanced Zoology*, 15:79-85.

Ambrose, D.P. and Livingstone, D. 1978. Population dynamics of three species of reduviids from peninsular *India. Bulletin entomology*, 19:201-203.

Ambrose D.P. and Livingston, D., 1985. Impact of mating on adult longivity oviposition pattern and incubation period on *Rhynocoris mariginatus, Environmental Ecology*, 3:99-102.

Ambrose, D. P., and Livingstone, D. 1986b. Bio-ecology of *Rhynocoris fusicpes* (Fab.) (Reduviidae) a potential predator of inset pests. *Uttar Pradesh Journl of Zoology*, 6: 36-39.

Ambrose, D.P. and Livingstone, D. 1988. Polymorphism in *Rhynocoris marginatus* Fabricius (Insecta: Heteroptera: Reduviidae). *Mitt. Zool. Mus. Berl,* 64: 343-348.

Ambrose, D. P. and Livingstone, D. 1989. Biology of the predaceous bug *Rhynocoris marginatus* Fabr. (Insecta: Heteroptera: Reduviidae). *Journal of Bombay Natural History and Society*, 86: 388- 395.

Ambrose, D.P. and Rani, M.R.S. 1991. Prey influence on the laboratory mass rearing of *Rhynocoris kumarii* (Ambrose and Livingstone) a potential biological control agent (Insecta: Heteroptera: Reduviidae). *Mitt. Zoo. Mes. Berl*, 67: 339-349.

Ambrose, D. P., and Rajan, K. 1995. Population dynamics of nine species of reduviids (Insecta: Heteroptera: Reduviidae) in Nambigaipuram semiarid zone, Southern India. *Journal of Soil biology and Ecology*, 15(1): 72 - 81.

Ambrose, D.P. and Sahayaraj, K. 1996. Longterm functional response of the reduviid predator *Acanthaspis pedestris* Stal in relation to its prey, *Pectinophora gossypiella* Saunders density. *Hexapoda,* 8 (2): 77-84.

Ambrose D.P. Sekar, P.C. and Kumaraswami, N.S. 1990a. Effect of starvation on the development reproduction and size of assassin bug *Rhynocoris marginatus, Environment and ecology,* 8: 548-555.

Ambrose, D.P. Saju, T. and Sahayaraj, K. 1990b. Prey influence on the development, reproduction and size of assassin bug, *Rhynocoris marginatus. Environmental Ecology*, 8 (1): 280 - 287.

Ambrose D.P. Solomon, K., and Vennison, S.J. 1985a. Effect of competion, space, and starvation and predatory behavior of the bug *Rhynocoris marginatus, Environmental ecology,* 3:280-285.

Ambrose, D.P., Solomon, K. and Vennison, S.J. 1985b. Effect of competition, space and starvation on the predatory behaviour of the bug *Rhynocoris marginatus; Environmental Ecology*, 3:280-285.

Anderson, M. T., J. M. Kiesecker, D. P. Chivers, and Blaustein, A. R. 2001. The direct and indirect effects of temperature on a predator – prey relationship. *Canadian Journal of Zoology,* 79: 1834-1841.

Antony, M., Daniel, J., Kurian, C. and Pillai, G.B. 1979. Attempts in introduction and colonization of the exotic reduviid predator *Platymeris laevicollis* Distant for the biological suppression of the coconut rhinocerous beetle, *Oryctes rhinocerous. Proceedings of Plant. Crops Symposium,* 2: 445-454.

Applebaum, S.W. 1985. Biochemistry of digestion in comprehensive Insect physiology biochemistry and pharma-cology (ed Kerkut. and Gilbert, L: 1) Pergamen Press. Toronto. 4: 279 – 311.

Arnold, C.Y., 1959. The determination and significance of the least temperature in a linear heat unit system. *Proceeings of Agricultural Socience and Horticultural Science,* 74 : 430-445.

Augusti. N. and Cohen, A.C. 2000. *Lygus hesperus* and L. Linedares (Hemiptera: Miridae). Phytophages, Zoophages or Omnivores evidence of feeding adaptations suggested day the salivary and midgut digestive enzymes. *Entomological Science* 35:176-86.

Awan, M.,S., 1983. A convenient recipe for reaing a predacious bug, *Oechalia schellenbegi* Guerin - Menille (Hemiptera : Pentatomidae). *Pakistan Journal of Zoology,* 15: 217 -218.

Babu, A., Seenivasagam, R. and Karuppa-samy, C. 1995. Biological control resources in social forest stands; In Biological control of social forest and plantation crops insects, T.N. Ananthakrishnan (Ed.) (New Delhi: Oxford and IBH Publishing Co. Ltd.,), pp. 7-24.

Bahrndorff, S., Marien, J., Loeschke, V., Ellers, J., 2009. Dynamics of heat-induced thermal stress resistance and Hsp70 expression in the springtail, *Orchesella cincta.* Functional Ecology 23 (2), 233–239.

Baker JE. 1991. Purification and partial characterization of amylase allozmes from lesser grain borer, *Rhyzopertha dominica. Insect Biochemistry,* 21:303-311.

Bakthavatsalam, N., Singh, S.P., Pushpalatha, N.A., and Bhummanavar, B.S. 1995. Optimum temperature for short term storage of eggs of *Chrysoperla carnea* (Stephens) (Neuroptera: Chrysopidae) *Journal of Biological Ccontrol,* 9 (1): 45-46.

Balogun, R.A. 1972. Digestive Carbohydrases and nature of amylase in the gut of *Zonoccus variegabus. Journal of Bulettinl Eenomological Society Nigeria, 3.* 91-94.

Balogun, R.A., and Fisher, O. 1970. Studies on the digestive enzyme of the common, *Bufo regularis* (Bonlinger). *Comparative Biochemistry and Physiology,* 33: 813- 820.

Barnard, E.A., and Prosser, C.L. 1973. Comparative biochemistry and physiology of digestion. In (Comparative animal physiology 3rd Edition by Prosser C.L. Saunders), *Philadelphia,* pp. 133 - 163.

Beckett, S. J., and Marton, R. 2003. Mortality of *Rhycopertha dominica* (F.) (Coleoptera: Bostrychiidae) at grain temperatures managing from 50°C to 60°C obtained at different rate of heating in a spouted bed. *Journal of Stored Products Research,* 39: 313 - 332.

Beckett, S.J., Morton, R., and Darby, J.A. 1998. The mortality of *Rhycopertha dominica* (F.) (Coleoptera: Bostrychiidae) and *Sitophilus oryzae* (L.) (Coteoptera: Curculionidae) at moderate temperature. *Journal of Stored Products Research,* 34: 363 - 376.

Beenakkers, A.M., Van der Horst, D.J., and Van Marrewijk, W.J.A. 1985. Insect lipids and lipoproteins, and their role in physiological processes, *Program of Lipid Research,* 24 : 19 - 67.

Bernfield, J.E., 1955. Amylase α and β In: Methods in Enzymology, Vol. 1. (Eds) Colowick, S.P. and Kaplan, N.O, 149-158.

Bhatnagar, V.S., Sithanantham, S.S., Pawar, C.S. Jadhav, D., Rao, V.R. Reed, W. 1983. Conservation and augmentation of natural enemies with reference to integrated pest managements in chickpea (Cicer arietinum L.) and pigeonpea (Cajanus cajan L.) Millsp; Proc. In. (Workshop on Integrated pest control for grain legumes, Goiani a, oia's Brazil eds). pp. 157-180.

Biggs, D.R. and Mc Greego, P.G. 1994. Gut pH and amylase and protease activity in larvae of the New-*Zealand grues* grab (Costelytra zealanticus) (Coleptera: Scarbaeidae) as a basis for selecting inhibitors. *Insect Biochemistry and Molecular Biology*, 26: 69-75.

Bignell, D.E., Eggleton, P., Nunes, L. and Thomas, K.L. 1997. Termites as mediators of carbon fluxes in tropical forest budgets for carbon dioxide and methane emission. In: Forest and Insects, Watt, A.D., Stork, N.E. and Hunter, M.D. (eds.). *Chapman and Hall Publication, London. United Kingdom,* PP. 109-234.

Boyer, P.D., Lardy, H. and Myrback, K., 1960. The enzymes. *Academic Press, New York*, 4:125-132.

Broderick, N.A., Courtney J Robinson, Matthew D McMahon, Jonathan Holt, Jo Handelsman3 and Kenneth F Raffa. 2009. Contributions of gut bacteria to *Bacillus thuringiensis*-induced mortality vary across a range of Lepidoptera. *BMC Biology,* 7 (11): doi:10.1186/1741-7007-7-11

Bradford, M.M. 1976. A rapid and sensitive method for the quantification of microgram quantities of protein utilizing the principle of protein-dye binding. *Analytical Biochemistry*, 72: 248–54.

Braga M.V., Pinto Z.T. and Lima M.M. 1998. Life cycle and reproductive pattern of Triatoma rubrofasciata (De Geer. 1773) (Hemiptera: Reduviidae), under laboratory conditions. *Members do instituto Oswaldo Cruz*, 93: 539-542.

Bragdon, T.H. 1957. Calorimetric determinations of blood lipids. *Journal of Biochemistry,* 190-513.

Braman, S. K., and Pendley, A. F. 1993. Temperature, photoperiod, and aggregation effects on development, diapause, reproduction, and survival in *Corythucha cydoniae* (Heteroptera: Tingidae). *Journal of Economic Entomology,* 28(4): 417-426

Braman, S. K, A. F. Pandley, B. Sparks, and Hudson, W. G. 1992. Thermal requirements for development, population trends, and parasitism of *Azalea lace* bug (Heteroptera: Tingidae). *Journal of Economic Entomology,* 85(3): 870-877.

Brauman, A., Bignell, D.E. and Tayasu, I. 2000. Soil feeding termites biology, microbial association and digestive mechanisms. In: Termites evolution, sociality, symbiosis, ecology.Abe, T. Bignell, D.E. and Higashi, M. (eds.). Klower Academic publishers, Dordrecht. Netherland. pp. 259.

Brauman, A., Dore, J., Eggleton, P., Bignell, D., Breznak, J.A. and Kane, M.D. 2001. Molecular phylogenic profiling of prokaryotic communities in guts of termites with different feeding habits. *FEMS Microbiology and Ecology*, 35: 27-36.

Breniere, S.F., Bosseno, M.F., Telleria, J., Carrasco, R., Vargas, F., Yaksic, and N., Noireau, F. 1995. Field application of polymerase chain reaction diagenosis and strain typing of *Trypanosoma cruzi* in Bolivian triatomines. *American Journal of Tropical Medical research,* 53: 179-184.

Breznak, J.A. and Brune, A. 1994. Role of microorganisms in the digestion of lignocellulose by termites. *Annual Review of Entomology,* 39: 453-487.

Breniere, J.F. Briere, P., Pracors, A.Y, Roux,L.E., and Priere, J.S. 1992. A novel rate model of temperature dependent development for arthropods, *Environmental Entomology,* 28: 22- 29.

Breniere,J., Pracros,P., Le Roux, A., and Pierre, J. 1999. A novel model of temperature –dependent development for arthropods. *Environmental Entomol,* 8:22-29.

Briggitte, B., Simone,F.B. 1998. Random by Amplified Polymorphic DNA Analysis of Sylvatic *Trypanosoma cruzi* Isolates Infers from French Guiana Accurate Phylogeny. Mem. Inst. *Oswaldo Cruz, Rio de Janeiro.,* 93(4): 485-486.

Broderick, N.A., Raffa, K.F., Goodman, R.M. and Handelsman, J.O. 2004. Census of the bacterial community of the gypsy moth larval midgut by using culturing and culture independent methods. *Applied and Environmental Microbiology,* 70(1): 293-300.

Brookes, V.J. 1969. The induction of yolk protein synthesis in the fat body of an insect. *Leucophala mederae,* by an analogy of the *Juvenile hormone, Developmental Biology,* 20: 459 - 471.

Brooks, M.A. 1963. The microorganisms of healthy insects. In: Steinhaus, E.A. (ed.). Insect pathology- An advanced treatise. *Academic Press, London,* pp. 250.

Buchner, P. 1965. Endosymbiosis of animals with plant microorganisms. Interscience, New York, N.Y.pp.103.

Buchanon, R.E. and Gibbons, N.E. 1979. Bergey's manual of determinative bacteriology. 8th Edition Williams and Wilkins, Baltimore, Maryland. pp. 269.

Buxton, P.A. 1930. The biology of a blood suckingbug, *Rhodnius prolixus Trans. Rent. Science London,* 78: 227-236.

Caceres, C., Ramirez, E., Wornoayporn, V. S., Mohammad Islam, S., Ahmad, S. 2007. A protocol for storage and long-distance shipment of mediternanean fruit fly (Diptera: Tephritidae) Eggs. Effect of temperature embryo age storage time on survival and quality. *Florida Entomology,* 90(1):110-114.

Campbell, A., Frazer, B.D., Gilbert, N., Gutierrez, A.P. and Mackauer, M. 1974. Temperature requirements of some aphids and their parasites. *Journal of Applied Ecology,* 11: 431–438.

Canals M., Soils R., Valderas J., Ehrenfeld M., Cattan PE 1997. Preliminary studies on temperature selection and activity cylces *Triatoma infestans* and *T. spinolai* (Heteroptera: Reduviidae) Chilean vectors of Chagas disease. *Journal of Medical entomology,* 34: 11-17.

Cappucino, J.G, and Sherman,N. 1999. Microbiology Laboratory Manual, 4th edition, Addision Wesley, England. PP: 1-477.

Carezza Bootto, M., Silvia, O., Marlene, R., and Pedro, E.C. 2005. dna evidence of *Trypanosoma cruzi* in the Chilean vector. *Mepraria spinolai* (Hemiptera: Reduviidae). . Mem. Inst. *Oswaldo Cruz, Rio de Janeiro,* 100(3): 236-239.

Carlos, G.S., Gildow, F.E., Fleisher, S.J., Cox-Foster, D., Lukszio, F.L. 2007. ELISA versus Immunolocalization to Determine the Assoication of Erwinia tracheiphila in *Acalymma vittatum* (Coleoptera: Chrysomelidae). *Evironmental Entomology,* 29(3): 542 –550.

Carroll A. and Quiring D. T. 1993. Interactions between size and temperature influence fecundity and longevity of a tortricid moth, *Zeiraphera canadensis.* Oecologia 93, 233–241.

Chapman, R.C. 1998. The insects: Structural and Function, Cambridge University Press, Cambridge. *Entomological Society of America, Lanham, MD.* pp. 1-17.

Chapman, R.F. 2000. The insects structure and function (4 th edition) Campridge *University Press, Campridge U.K.* pp. 112-120

Chaterjee, G.K., Rao, G.P., Aswath, S.K., and Chatterjee, S.N. 1989. Studies on the protease activity in the digestive juice of different breads / race of mulberry silkworm *Bombyx mori. Newsletter,* 4(3) : 6 -7.

Cheong, Y.L., Sajap, A.S. Hafidzi, M.N., Omar, D. and Abood, F. 2010. Outbreaks of bagworms and their natural enemies in an oil palm, *Elaeis guinmeensis,* plantation at Hutan Melintang, perk, Malaysia. *Journal of Entomology,* 7(3): 141 – 151.

Chen, D., and Purcell, A. 1997. Occurrence and transmission of facultative endosymbionts in aphids. *Current Microbiology,* 34: 220-225.

Cherian, M.S. and Brahmachari, L. 1941. Notes on three predatory hemipteran from South India, *Indian Journal of Entomology.,* 3:115-118.

Cherian, M.C. and Kylasam, M.S. 1939. On the biology and feeding habits of *Rhynocoris fuscipes* (Fab.) (Heteroptera: Reduviidae). *Journal of Bombay Natural History, Society*, 61: 256-259.

Christian L., Jaceques F. and Jorg, E G. 1999. Development of *Rhodnius prolixus* (Hemiptera: Reduviidae) under constant and cyclic conditions of Temperature and Humidity. Mem. Inst. Oswaldo Guz. 3 *Rio de Janeiro*, 94: 403-409.

Clark, N. 1996. The effect of temperature and humidity upon eggs of the bug, *Rhodnius prolixus* (Heteroptera: Reduviidae). *Journal of Animal Ecol*ogy, 4: 82-87

Clark, L.R., Gejer, R. W., Huges, R.W., and Morris, R.F. 1978. The ecology of Insect Population in theory and practice. English languages Book Society. Chapman and Hall, ElBS PP. 26-56.

Claver, M.A. and Ambrose, D.P. 2001. Evaluation of *Rhynocoris kumarii* Ambrose & Livingstone (Hemiptera: Reduviidae) as a potential predator of some lepidopteran pests of cotton. *Journal of Biological Control*, 15(1): 15-20.

Claver, M.A. and Ambrose, D.P. 2001. Suitability of substrata for the Mass rearing *of Rhynocoris fuscipes* (Heteroptera: Reduviidae), a key predator of pod borer *Helicoverpa armigera* (Hubn.). *Entomon*, 26(2): 141-146.

Claver, M.A. and Ambrose, D.P. 2003. Influence of mulching and intercropping on the abundance of the Reduviid predator, *Rhynocoris fuscipes* (F.). *Biol.control of insect pests,* Ignacimuthu, S. and Jayaraj, S. (eds.). New Delhi, India: Phoenix Pub. House Pvt. Ltd., p.183-191.

Claver, M.A. and Ambrose, D.P. 2003. Suppression of *Helicoverpa armigera* (Hubner), *Nezara viridula* (L.) and *Riptortus clavatus* Thunberg infesting pigeonpea by the reduviid predator *Rhynocoris fuscipes* (Fabricius) in field cages. *Entomologia Croatica,* 7 (1-2): 85-88.

Cohen, A.C. 1982. Water and temperature relations of two hemipteran members of a predator – prey complex. *Envionmental Entomology,* 11: 715-719.

Cohen, A.C. 1993. Organization of digestion and preliminary characterization of salivary trypsin-like enzymes in predaceous heteropteran, *Zelus renardii. Journal Insect Physiology,* 39(10): 823-829.

Cohen, A.C. 1998. Biochemical and morphological dynamicsand predatory feeding habits in terrestrial Heterop-tera, In. J. R. Ruberson and M. Coll (eds.), Predaceous Heteroptera: implications for biological control. Thomas say publication Entomology. *Entomological society of America, Lanham, MD.* pp. 21-32

Cohen, A.C. 2000. How carnivorous bugs feed,. In C.W. Schaefer and A.R. Panizzi (eds.), Heteroptera of economic importance. *CRC, Boca Raton, FL.* pp. 563-570.

Colourick, K.G., and Kaplan, C. 1959. Methods in Enzymology. Vol. 6 *Academic Press New York*.pp; 143-147.

Crocker, A. 1975. Components of the feeding niches of *Geocoris* spp. (Hemiptera: Lygaeidae). Ph.d Disser-tation, University of Florida, Gainseville, USA.

Das, S. S.M. 1996. Biology and behaviour of chosen predatory hemipterans (Insecta: Heteroptera). Ph.D. thesis, Manonmanium Sundaranar University, Tirunvelvli, India.

Das, S. S.M. and Ambrose, D.P. 2008. Redescription, biology and behaviour of a harpactorine assassin bug *Vesbius sanguinosus* STÅL (Insecta, Hemiptera, Reduviidae). *Polish Journal of Entomology,* 77: 11-29.

Das, S. S.M. and Ambrose, D.P. 2008. Redescription, biology and behaviour of a harpactorine assassin bug *Irantha armipes* (Stål) (Insecta: Hemiptera: Reduviidae). *Acta Entomologica Slovenica,* 16(1): 37-56.

Dasch, G. A., E. Weiss, and Chang, K. P. 1984. Endosymbionts of insects,. In N. R. Krieg and J. G. Holt (ed.), Bergey's manual of systematic bacteriology, vol. 1. *Williams & Wilkins, Baltimore, Md.* pp.811–833.

Das, H.K. 2005. A Text book of Biotechnology. A 2nd edition willey publing Inc., USA. Page NO.1000-1015.

David, P.M., and Natarajan, S. 1989. The Hindu June 21, pp-24.

De Bach, P., and Hagen, K.S. 1964. Manipulation and entomophages species in Biological control of insects pests and weeds, (ed) Paul De Bach (New York) Rembold publications corporation, pp-429-458.

De Bach, P., and Hagen, D. 1991. Biological control by natural enemies 2nd ed. *Cambridge University Press, Cambridge.*

De Clercq, P., and Degheele, D. 1992a. A meat-based diet for rearing the predatory *stinkbugs* Podisus maculiventris and *Podisus sagitta* [Hetroptera: Pentatomidae]. *Entomophaga,* 37 (1): 149 –157.

De Clercq., P., and Degheele, D. 1992b. Development and survival of *Podisus maculiventris* (Say) *Podisus sagitta* (Fab.) (Heteroptera : pentatomidae) at various constant temperatures. *Canadian Entomology,* 124: 125-133.

De clercq, P., and Degheel, D. 1993. Cold storage of the predatory bugs *Podisus maculivetris* (say) and *Podisus sagitta* (Fabricius) (Heteroptera: Pentatomidae). *Parasitica,* 49: 27-41.

Denlinger, D.L. 1991. Relationship behaven cold hardines and diapaure, In R.E. Lec and D.L. Denlinger, eds. Insects at low Temperature. *New York Chapman and Hall,* pp 174 - 198.

Dhanasing and Ambrose, D.P. 2006. Seasonal density of reduviids predators of Vagaigulam Semi arid ecosystem in Thoothukudi district Tamilnadu. *Insect Environment,* 12 (1):24-25.

Dillon, R.J. and Dillon, V.M. 2004. The gut bacteria of insects: Non-pathogenic interactions. *Annual Review of Entomology,* 49: 1-16.

Diodonet, J., Zaninscio, J.C., Sidiyama. C.S., and Picanco, M.C. 1996. *Desenvolviments sobreviencia* nifal de *Podisus nigrispinus* (Dallas) e de *Supputius cinctipes* stal (Heteroptera: Pentatomidae) em differentes temperature. *Journal Brazilian Zoology,* 12: 513-518.

Dodd, C.S. 2004. Development and optimization of PCR based techniques in predator gut analysis. Ph.D. Thesis, Cardiff University, Cardiff.

Donovan, S.E., Purdy, K.J., Kane, M.D. and Eggleton, P. 2004. Comparison of Euarchae strains in the guts and food soil of the soil feeding termite *Cubitermes fungifaber* across different soil types. *Applied and Environmental Microbiology,* 70(7): 3884-3892.

Dooremalen, C. and Ellers, J. 2010. A moderate change in temperature induces changes in fatty acid composition of storage and membrane lipids in a soil arthropod. Journal of Insect Physiology 56 : 178–184.

Douglas, A.E. 1992. Microbial brokers of insect-plant interactions. Proceedings of 88th International symposium on Insect-plant relationships, Dordecht, Neth, Kluwer. pp. 329-336.

Dunborn, D.M., and Bacon, O.G. 1972. Influence of temperature on development and reproduction of *Geocoris atricolor, Geocoris pallens,* and *Geocoris punctipes* (Heteroptera: Lygaeidae) from California. *Enviornmental Entomology,* 1: 596 - 599.

Durbin, E.G., Casas, M.C., Rynearson, T.A., Smith, D.C., 2008. Measurement of copepod predation on nauplii using qPCR of the cytochrome oxidase I gene. Marine Biology 153, 699–707.

Eckert, M., Agricola, H., and Penzlin, H. 1981. Immunocyto chemical identification of proctolinklibe immunbo reactivity in the termional ganglion and hindgut of the cockroach periplannata Americana (L.) *Cell Tissue Research,* 217: 633 – 645.

Edwards, J.S. 1962. Observations on the development and predatory habits of two reduviids (Heteroptera), *Rhynocoris carmelita* Stal and *Platymeris rhadamanthus* (Gerst). *Proceeding of Research Entomo-logical Society of London,* (A) 37: 89-98.

Eguchi, M., and Iwamoto, A. 1976. Alkaline protease, in the midgut tissue of and digestive fluid of the silkworm *Bombyx mori. Insect Biochem*istry, 6: 491-496.

Eguchi, M. M.1983. Relationship between alkaline proteases from the midgut lumen and epithelia of the silkworm: Solubilisation and activation of epithelial protease (6B3). *Comparative Biochemistry and Physiology,* 75: 589 - 593.

Ehler, L. E., Long, R. E., Kinsey, M. G. and Kelley, S.K. 1997. Potential for augmentative biological control of black bean aphid in California sugarbeet. *Entomophaga,* 42 (I/2) : 241-256.

Ehrlich, H.A. 1989. PCR Technology. Principles and Applications for DNA Amplification Stockton Press, New York.pp.67-69.

El-Wakeil, N.M.E. 2003. New aspects of biological control of *Helicoverpa armigera* in organic cotton production Ph.D Dissertation. Institute of plant pathology and plant protection, George – August University, Gothingen, German.

Engelman, F. 1979. Insect vitellogenin: Identification, biosynthesis and role in vitellogenesis. *Advanced Insect Physiology,* 14: 49-108.

Engvall, E., and Perlman, P. 1971a. Enzyme linked immunosorbent assay, ELISA. III. Quantitation of specific antibodies by enzyme – labeled anti-immunoglobulin in antigen coated tubes. *Journal of Immunology,* 109: 129-135.

Engvall, E., and Perlman, P. 1971b. Enzyme linked immunosorbent assay, (ELISA), Quantitative assay of immunoglobulin G. *Immunochemistry,* 8: 871 – 874.

Fadare, T.A. and Amusa, N.A. 2003. Comparative efficacy of microbial and chemical insecticides on four major lepidopterous pests of cotton and their (insect) natural enemies. African Journal of Biotechnology, 2 (11): 425-428

Ferro, D.N., and Chapman, R.B. 1979. An effect of different constant humidifies and temperature on the spotted spider mite egg hatch. *Environmental Entomology,* 8: 701-705.

Fichter, B. L., and Stephen, W. P. 1981. Time related decay in prey antigens ingested by the predator *Podisus maculiventris* (Hemiptera: Pentatomidae) as detected by ELISA. *Oecologia,* 51: 404–407.

Frazer, B.D. and Mc Gregor, R.R. 1992. Temperature dependent survival and hatching rate of eggs of seven species of coccinellidae. *Canadian Entomology,* 124: 305 - 218.

Forbes, S.A. 1983. The food relations of the Carabidae and Coccinelidae. *Bulletin of the Illinois State laboratory of Natural History,* 1: 33-64.

Fukatsu, T. and Hosokawa, T. 2002. Capsule transmitted gut symbiotic bacterium of the Japanese common plataspid stinkbug, *Megacopta punctatissima. Applied and Environmental Microbiology,* 68(1): 389-396.

Garcia, M.R., Montilla, M., Nicholls, S., and Alvaez, D. 2002. Population Genetic Analysis of Colombian *Trypanosoma cruzi* Isolates Revealed by Enzyme Electrophoretic Profiles. *Memories of Insect Oswaldo Cruz, Rio de Janeiro,* 96 (1) 31-51.

Garcia, S.L., Rodrigues, V.L., Garcia, N.L., Ferraz filho, A.N., and Mello, M.L.S. 1999. Survival and molting incidene after heat and cold schocks in *Panstrongylus megistus* (Burmeister) Mem. Ins. Oswaldo cruz 94: 131-136.

Garcia-Salazar, C.F.E., Gillow.. S.J., Fleischer, D. Coxfoster, and Luezie, F.L. 2000. Alimentary canal of adult Acalymma vittata (F.) (Coleoptera: Chrysomelidae). Morphology and potential role in the survival of *Erwinia trachiphila* (Enterobacteriaceae). *Canadian Entomology,* 132:1-3.

Garnoel N., and Barett A.C. 1993. Characterization of differences between whiteflies using RAPD – PCR. *Insect Molecular Biology,* 2: 33-38.

George, P.J.E., and Ambrose, D.P. 1998. Relative toxicity of five insecticides to *Rhynocoris fuscipes* (Fab.) as a potential predator of insect pest (Insecta: Heteroptera: Reduviidae). *Shaspha,* 5(2):197-202.

George, P. J. E., and Ambrose, D. P. 1999a. Biochemical modulations by insecticides in a non-target harpactorine reduviid *Rhynocoris kumarii* Ambrose and Livingstone (Heteroptera: Reduviidae). *Entomon,* 24(1): 61 - 66.

George, P.J.E., and Ambrose, D.P. 1999b. Impact of insecticides on the biochemical constituvents in nontarget harpactorinae Reduviids *Rhynocris fuscipes* (Fab.) *Shaspha,* 6(2): 167-172.

George, P.J.E., and Ambrose, D.P. 1999c. Post embryonic developmental changes in nontarget *Rhynocris fuscipes* (Fab.) Insecta: Heteroptera: Reduviidae. *Indian journal of Environmental Science,* 3(1): 201-206.

George, P. Claver, J.E., and Ambrose, D.P. 2000. Life table of *Rhynocoris fuscipes* (Fabricius) (Heteroptera: Reduviidae) reared on *Corcyra cephalonica* (Stainton); *Pest Management and Economic Zoology,* 8(1): 57 – 63.

George, P. J. E. and Ambrose, D. P. 2000b. Nymphal cannibalism in reduviids a constraint in mass rearing. Biotechnological applications for integrated pest management (S. Ignacimuthu, A. Sen and S. Janarthanan eds.), Oxford and IBH Publishing Co. Pvt. Ltd., New Delhi, India.

George, P. J. E., Kannagi, J. and Ambrose, D. P. 2002. Nutritional influence of prey on the biology and biochemistry in *Rhynocoris marginatus* (Fabricius) (Heteroptera: Reduviidae). *Journal of Biological Control,* 16(1): 1 - 4.

Gilchrist, G. W. 1995. Specialist and generalists in changing environments, fitness landscapes of thermal sensitivity. *The American Naturalist,* 146(2): 252-270.

Goel S.C. 1978. Bioclogical studies of two years capture of Hemipteran in Westeran Uttar pradesh Oriental Ins.*The Insect World,* 12: 369 – 376.

Gomez – Nunez, J.C., and Fernandez, J..M. 1963. La colonia de *Rhodnius prolterus* en el Instituto vonezoland de Investigacions cientificas. *Bol. Inf Dir Mal San Amb,* 3: 132– 137.

Gopalakrishnan, C. 2001. Mass production and utilization of microbial agents with special reference to insect pathogens. In: Ignachimuthu, S. and Sen, A. (eds.). Micorbials in insect pest management. Oxford and IBH Publishing Co. Pvt. Ltd., New Delhi. pp. 174.

Goryshin, N.I., Tuganova I.A. 1989. Optimization of short tem storage of eggs of the predatory bug *Podisus maculventris* (Hemiptera: pentatomidae), *Zoological – Cheskii zhurnal,* 68 : 111-119.

Gozlan, S., Millot, P., Rousset, A. and Fournter, D. 1997. Test of RAPD-PCR method to evaluate theve efficiency of augmentative biological control with *Orius* (Heteroptera: Anthocoridae). *Entomophaga,* 42: 593-600

Greenstone, M. H. and Trowell, S. C. 1994. Arthropod predation—A simplified immunodot format for predator gut analysis. *Annual Entomological Society of America* 87: 214–217.

Greenstone, M. H. and Hunt, J. H. 1993. Determination of prey antigen half-life in *Polistes metricus* using a monoclonal antibodybased immuno-dot assay. *Entomologia Experimentalis Applicata,* 68, 1–7.

Greenstone, M.H., Rowby, D.L., Heimbach, U., Landgren, J.G., Pamenstiel, R.S. and Rchner, S.A. 2005. Barcoding generalist predators by polymerase chain reaction: Carabids and Spiders. *Molecular Ecology,* 14: 3247-3266.

Greenstone, M.H., Rowley, D.L., Weber, D.C., Payton, M.E., Hawthorne, D.J., 2007. Feeding mode and prey detectability half-lives in molecular gut-content analysis: an example with two predators of the Colorado potato beetle. Bulletin of Entomological Research 97, 201–209.

Grundy PR. 2004. Integration of the assassin bug *Pristhesancus plagipennis* (Walker) (Hemiptera: Reduviidae) as a biological control within an Integrated Pest Management program for cotton. *Australian Journal of Entomology* 43, 77-82.

Grundy, P. 2007. Utilizing the assassin bug, *Pristhesancus plagipennis* (Hemiptera: Reduviidae), as a biological control agent within an integrated pest management programme for *Helicoverpa* spp. (Lepidoptera: Noctuidae) and *Creontiades* spp. (Hemiptera: Miridae) in cotton. *Bulletin of Entomological Research,* 97: 281 - 290.

Grundy, P. and Maelzer, D. 2000. Assessment of *Pristhesancus plagipennis* (Walker) (Hemiptera : Reduviidae) as an augmented biological control in cotton and soybean crops. *Australian Journal of Entomology,* 39: 305-309.

Grundy PR, Maelzer DA, Collins PJ & Hassan E. 2000. Potential for integrating eleven agricultural insecticides with the predatory bug *Pristhesancus plagipennis* (Hemiptera: Reduviidae). *Journal of Economic Entomology* 93, 584-589.

Grundy PR & Maelzer DA 2002a. Augmentation of the assassin bug *Pristhesancus plagipennis* Walker (Hemiptera: Reduviidae) as a biological control agent for *Helicoverpa* spp. in cotton. *Australian Journal of Entomology* 41, 192-196.

Grundy PR & Maelzer DA 2002b. Factors affecting the establishment and dispersal of *Pristhesancus plagipennis* Walker(Hemiptera: Reduviidae) nymphs when released onto soybean, cotton and sunflower crops. *Australian Journal of Entomology* 41:272-278.

Gwynne, D.T. 1989. Does the copulation incrase the risk of Predation. TREE 4 (2): 54 – 56.

Hagerty, A.M., Mcpherson, J.E., and Bradshaw J.D. 2001. Life history and laboratory rearing of *Emesaya brevipennis* (Heteroptera: Reduviidae) in Southern illionis. *Florida Entomologist,* 83(1): 58-63.

Haglar, J.R., and Naranjo, S.E. 2004. A multiple ELISA system for simultaneously monitoring intercrop movement and feeding activity of mass-released insect predators. *International Journal of pest management,* 50: 199-207

Haglar, J.R., Naranjo, S.E., Bradley – Dunlop – D., Enriquez, F.J., and Henneberry, T.J. 1994. A monoclonal antibody to pink bollworm (Lepidoptera: Gelechidae) egg antigen – a tool for predator gut analysis. *Annals of the Entomological Society of America,* 87: 85-90.

Hagler, J. R., Cohen, A. C., Bradley-Dunlop, D., and Enriquez, F. J. 1992. Field evaluation of predation on *Lygus hesperus* using a species- and stage-specific monoclonal antibody. *Environental Entomology,* 21: 896–900.

Hagler, J.R. and Cohen, A.C. 1990. Effects of time and temperature on digestion of purified antigenby Geocoris punctipes (Hemiptera: Lygaeidae) reared on artificial diet. *Ann. Entomol. Entomo.Exp.Appl,* 68, 1-7.

Hagler, J.R. and Naranjo, S. 1997. Measuring the sensitivity of an Indirect predator gut content ELISA detectability of prey remains in relation to predator species temperature time, and meal size. *Biological Control,* 9, 112-119.

Hagler, J.R. and Naranjo, S.E. 1994a. Determining the frequency of heteropteran predation on sweet-potato whitefly and pink-bollworm using multiple ELISAs. *Entomologia Experimentalis et Applicata,* 72: 59–66.

Hagler, J.R. and Naranjo, S.E. 1994b. Qualitative survey of two coleopteran predators of Bemisia tabaci (Homoptera: Aleyrodidae) and *Pectinophora gossypiella* (Lepidoptera: Gelechiidae) using multiple prey gut content ELISA. *Environmental Entomology,* 23: 193–197.

Hagler, J.R. and Naranjo, S.E. 1999a. Determining the frequency of heteropteran predation on sweet-potato whitefly and pink-bollworm using multiple ELISAs. *Entomologia Experimentalis et Applicata,* 72: 59–66.

Hagler, J.R. and Naranjo, S.E. 1999b. Qualitative survey of two coleopteran predators of Bemisia tabaci (Homoptera: Aleyrodidae) and *Pectinophora gossypiella* (Lepidoptera: Gelechiidae) using multiple prey gut content ELISA. *Environmental Entomology,* 23: 193–197.

Haines, C. P. 1991. Insects and Arachnids of Tropical Stored Products: Their Biology and Identification (A *Training Manual).* 2nd ed. Natural Resources Institute, Chatham, UK, pp. 75–76.

Haridass E.T. 1987. Reproduction in some Reduviidae from southern India (Heteroptera : Insects). In: (Advance Biological control Research in India) K.J. Joseph and UC Abdurahiman, Calicut(eds.) : Printex India pp. 56-64.

Haridass, ET., and Ananthakrishnan, T.N. 1981. Functional morphology of the salivary system in some Reduviid (Insecta: Heteroptera). *Proc. Indian Acad. Sci,* 90: 145-60.

Harklein, P. 2003. Microbiology. International edition Mc Grahill. Higher Education – 1009.pp. 112-120.

Harper, G.L., King, R.A., Dodd, C.S., Harwood, J.D., Glen, D.M., Bruford, M.W. and Symondson, W.O.C. 2005. Rapid screening of invertebrate predators for multiple prey DNA targets. *Molecular Ecology,* 14: 819–828.

Harwood, J.D, and Obrycki, J.J. 2005a. Quantifying aphid predation rates of generalist predators in the field. *European Journal of Entomology,* 102: 335–350.

Harwood, J.D., and Obrycki, J.J. 2005b. Quantifying aphid predation rates of generalist predator for multiple prey DNA target. *Molecular Ecology,* 14: 819-828.

Harwood, J.D., Desneux, N., Yoo, H.J.S., Rowley, D.L., Greenstone, M.H., Obrycki, J.J., O' Neil, R.J., 2007. Tracking the role of alternative prey in soybean aphid predation by Orius insidiosus: a molecular approach. Molecular Ecology 16, 4390–4400.

Harwood, J.D., Phillips, S.W., Sunderland, K.D., and Symondson, W.O.C. 2001. Secondary predation: quantification of food chain errors in an Aphid-Spider-Carabid system using monoclonal antibodies. *Molecular Ecology,* 10: 2049-2057.

Harwood, J.D., Sunderland, K.D. and Symondson, W.O.C. 2004. Prey selection by linyphiid spiders: molecular tracking of the effects of alternative prey on rates of aphid consumption in the field. *Molecular Ecology,* 13: 3549–3560.

Harwood, J.D., Yoo, H.J.S., Rowley, D.L., Greenstone, M.H., O' Neil, R.J., 2009. Differential impact of adults and nymphs of a generalist predator on an exotic invasive pest demonstrated by molecular gut-content analysis. Biological Invasions 11, 895–903.

Helosia, S.L., Coelho, Georgia C., Atella, Monica, F., Moreira, K., Gondim, C., and Hatisaburo, M. 1997. Lipophorin density variation during ogenesis in *Rhodnius prolixus. Insect Biochemistry and Physiology,* 35: 301 -313.

Herrera, C.J., Van Driesche, R.G., and Bellotti, A.C. 2005. Temperature – dependent growth rates for the casava mealybug, *Phenacoccus herreni,* and two Encyritid parasitoids, *Epidinocasysis diversicornis* and *Acerophages coccois* in colombia. *Entomologia Experimentails et Applicata,* 50: 21-7.

Hill, D. S. 1990. *Pests of Stored Products and Their Control.* Belhaven Press, London. 69 pp.

Hiramath, I.G. and Thontadarya, T.S. 1983. Natural enemies of sorgam earhead bug *Calocoris aggustatus lethiry* (Hemiptera: Miridae): *Currrent Research,* 12: 10-11.

Hochachka, P.W., Somero, G.N., 2002. Biochemical Adaptation: Mechanism and Process in Physiological Evolution. Oxford University Press, Oxford, 560 pp.

Hoogendoorn, M. and Heimpel, G.E. 2001. PCR-based gut content analysis of insect predators: using ribosomal ITS- 1 fragments from prey to estimate predation frequency. *Molecular Ecology,* 10: 2059–2067.

House, H.L. 1965. Digestion in the physiology of Insecta (Ed by Rockstein,M) *Accademic press Newyork,* 2: 815-852.

Hoy, M. A. 1994. Insect Molecular Genetics: An Introduction to Principles and Applications. Academic Press, San Diego, California, U.S.A.

Hunt, J. and Charnley, A.K. 1981. Abundance and distribution of the gut flora of the desert locust, *Schistocerca gregaria. Journal of Inventebrate Pathology,* 38: 378-385.

Imms, A.D. 1985. A general text book of Entomology. (London: The English Language Book (Society and Methuen Company Limited). Pp 459-460.

Isenbour, D.J., and Yeargan, K. V. 1981. Effect of temperature on the development of *Orius insidiosus*, with notes of laboratory rearing. *Annual Entomological Society of America*, 74: 114 - 116.

Ishaaya, I. E. and Swirski, E. 1970. Invertase and amylase activity in the armoured scales *Chrysomphabes aonidium* and *Anonidicella auranti*. *Journal Insect Physiology*, 16: 1599 – 1606.

Ishaaya I, Moore I, and Joseph, B.1971. Protease and amylase activity in the larvae of the Egyprioan cotton worm, *Spodoptera littoralis*. *Journal of Insect Physiology*, 17:945-953.

Islam, S.S. and Chapman, R.B. 2001. Effect of temperature on predatonby Tasmanian lacewing larvae. *New Zealand Plant Protection*, 54:244-247.

Ito, K. and Nakata,T. 2000. Geographical variation of Photoperiodic response in the females of a predatory bug, *Orius sauteri* (Poppius) (Heteroptera: Anthocoridae) from Northern Japan. *Applied Entomology and Zoology*, 35: 101-105.

Izumi, and Ohto, 2001. "Effect of temperature of *Orius strigicolius* (Heteroptera: Anthocoridae) fed on *Frankliniella occidentalis* (Thysanoptera: Thripidae)". *Applied Entomology and Zoology*, 36(4): 483 - 488.

Jalali, M. A., Tirry, L. and Patrick De Clercq, 2010. Effect of temperature on the functional response of Adalia bipunctata to Myzus persicae. BioControl, 55:261–269 (DOI 10.1007/s10526-009-9237-6)

James, D.G. 1992. Effect of temperature on development and survival of *Pristpesancus plagipennis* (Fab.) (Hemiptera: Reduviidae) *Entomo-phaga*, 37(2) : 259-264.

James, D.G. and Vogele, B. 2001. The effect of imidacloprid on survival of some beneficial arthropods. *Plant Prot. Quart.* 16: 58–62.

Jaronski, S.T., Lord, J., Rosinska, J., Bradley, C., Hoelmer, K., Simmons, G., Osterlind, R., Brown, C., Staten, R. and Antilla, L. 1998. Effect of a *Beauveria bassiana*-based mycoinsecticide on beneficial insects under field conditions. In: Proceedings, Brighton Crop Protection Conference, Vol. 2. British Crop Protection Council, Brighton, UK, pp. 651–656.

Jeffrey, P.S. and Jesusa, C. L. 2006. Assesing Biochemical fitness of predator *Podisus maculiventris* (Heteroptera: Pentatomidae) in relation to Food quality: Effects of five species of prey. *Journal of Applied Entomology*, 4 (5): 24-26.

Joseph, M.T. 1959. Biology, bionomics and economic importance of reduviids collected from Delhi, *Indian Journall of Entomology*. 24: 46-48.

Jones, C.G. 1984. Microorganisms as mediatirs of plant resource of exploitation by insect herbivores. In: A new Ecology: Novel Approaches to Interactive

Systems. (eds.p.w.price; Slobodchikoff, C.N and Gaud, W.S) Wiley and Sons, New York, pp. 53-99.

Jones, D., and Sterling W.L. 1979. Temperature thresholds for spring emergence and flight of the boll Weevil. *Environmental Entomology,* 8: 1118-1122.

Jones, P.A., and Coppland, H.C. 1963 Immature stages and biology of *Aproerema Carnea* (Hemiptera: Pentatomidae). *Canadian Entomology,* 95: 770-779.

Kalaiselvam, P.T. and Arul Pandi,I. 2006. Bioprocess of Technology. *MJP Publishers,* 278-298.

Kino, J.H., Kim, M.W., Han, G.S., and Lee, J.O. 1999. Effect of temperature on the development oviposition of minute pirate bug, *Orius slauteri* and *Orius minutus* (Heteroptera: Anthocoridae). *Applied Entomology and Zoology,* 32 : 644-648.

King, R.A., Read, D.S., Traugott, M., and Symondson W.O.C. 2008, Molecular analysis of predation, a review of best practice for DNA-based approaches. *Molecular Ecology,* 17: 947 – 963.

Knapp, R, and Casey, T.M. 1986. Thermal ecology, behavior and growth of gypsy moth and eastern tent caterpillars. Ecology 67 : 598-608.

Kohno, K., and Kashio,T. 1998. Thermal effects on reproductive diapause induction in *Orius Sauteri* (Heteroptera : Anthocoridae). *Appllied Entomological Zoology,* 33:487-490.

Kumaraswami, N. S. 1991. Bioecology and ethology of chosen predatory bugs and their potential in biological control. Ph. D. Thesis, Madurai Kamaraj University, Madurai, India.

Kumaraswami N. S., and Ambrose D.P. 1993 population dynamics of five assassin bugs from Melapattam scrub jungle of south India: *Hexapoda,* 23:12-16.

Kumaraswami N.S., and Ambrose D.P.1994 population dynamics of assassin bugs from the courtallam tropical evergreen forest in Western ghates in Tamilnadu; *Journal Bombay Natural. Society,* 91 : 260 – 267.

Kunkel, J.G. and Nordin, C. 1985. Yolk Proteins, In : Comprhensive Insect Physiology Biochemistry and Pharmacology Vol.I. Kergut, G.A. and Gilbert, L.(Editors). *Pergamon Press Ltd,: Oxford,* 33: 105-111.

Laemli, U.K. 1989. Cleavage of structural the assembly of the head of bacteriophage T4. *Nature,* 227: 680-685

Lakkundi, N. H. 1989. Assessment of reduviids for their predation and possibilities of their utilization in biological control. Ph. D. Thesis, IARI, New Delhi, India.

Lakkundi, N. H., and Parshad, B. 1987. A technique for mass multiplication of predator with sucking type of mouth parts with special reference to reduviids. *Journal of Soil Biology and Ecology*, 7: 65 - 69.

Law. J.A., Dunn,D.V. and Kramer. K.J. 1977. Insect proteases and peptidases. *Adv. Enzymology,* 45: 389 - 425.

Lee, R.E. 1991. Principles of insect low temperature tolerance. In "Insects at low Temperature" (R.E. Lee, Fr., and D.L., Denlinger, Eds), *Chapman and Hall, New York / London,* 17-46.

Lees, A.D. 1955. The physiology of diapause in Arthropods. *Camb. Monograph in Experimental Biology,* 41: 151.

Legaspi, J. C. (2004) Life history of *Podisus maculiventris* (Heteroptera: Pentatomidae) adult females under different constant temperatures, Environ. Entomol. 33 (5): 1200-1206

Lehman, R.M., Lundgren, J.G., Petzke, L.M., 2009. Bacterial communities associated with the digestive tract of the predatory ground beetle, Poecilus chalcites, and their response to laboratory rearing and antibiotic treatment. Microbial Ecology 57, 349–358.

Lemke, T., Stingl, U., Egert, M., Friedrich, W. and Brune, A. 2003. Physicochemical conditions and microbial activities in the highly alkaline gut of the humus feeding larvae of *Pachmoda ephippiata* (Coleoptera: Scarabacidae). *Applied and Environmental Microbiology,* 69(11): 6650-6658.

Lie, D.X., Y.L. Hou, and Shen, Z.R. 2005. Influence of host plant species on the development and reproduction of hawthorn spider mite. *Acta Ecology and Sinica,* 25: 1562–1568.

Livingstone, D. and Ambrose, D.P. 1978. Feeding behaviour and predatory efficiency of some reduviids from the Palghat gap India; *Journal of Madras University,* 41: 1-25.

Logan, J. D. Wolesenky, W. and Joern A. 2007, Insect development under predation risk, variable temperature, and variable food quality, *Maths Biostatistica and Engineering,* 4(1): 47–65.

Logan, J. D. Wolesenky, W., and Joern, A. 2006, Temperature-dependent phenology and predation in arthropod systems, *Ecological Modelling,* 196: 471–482.

Louis, D. 1974. Biology of Reduviidae of cocoa farms in Ghana. *The American Midland Naturalist,* 91: 68–89.

Lounibos, P. L., Martin, E.A., and Duzak, Escher .E.L. 1998. Daylength and temperature Control of predation, body size, and rate of increase in

Toxorhynchites rutilus (Diptera: Culicidae). *Annual Entomological Society of America,* 91(3): 308-314.

Lowry, O.H., Rosenbrough, J.J., Farr, A.L., and Randall, R.J. 1951.Protein measurement with   the folin phenol reagent, *Journal of Biology and Chemisry,* 193:263-275.

Luck, R. F., Shepard, B. M., and Kenmore, P. E. 1988. Experimental methods for evaluating arthropod natural enemies. *Annual Review of Entomology,* 33: 367–391.

Lundgren, J.G., 2009a. Relationships of Natural Enemies and Non-prey Foods. Springer International, Dordrecht, The Netherlands.

Lundgren, J.G., Lehman, R.M., Chee-Sanford, J., 2007. Bacterial communities within digestive tracts of ground beetles (Coleoptera: Carabidae). Annals of the Entomological Society of America 100, 275–282.

Luz, C., Fargnes, J., and Qrune wald, J. 1998. The effect of fluctuating temperature and humidity on the longevity of stared *Rhodnius prolixus* (Triatomine*). Journal Applied Entomology,* 122: 219 - 222.

Lysenko, O. 1985. Non – Spore forming bacteria pathogenic to insects. Incidence and mechanisms. *Annual Review of Microbiology,* 35(3): 235-242.

Ma, M., Burkholder, J.K., Webb, R.E., and Hsu, H.T. 1984. Springer Series in *Experimenal Entomology*, Printed in the united states, pp.44-47.

Maa,C.J.W.1987. Hydroprene induction of Carboxyl esterase on house-fly (*Musca domestica*). *Chinese Journal of Entomon,*7: 49-58.

Mahdian K, Vantornhout I, Tirry L, De Clercq P 2006. Effects of temperature on predation by the stinkbugs *Picromerus bidens* and *Podisus maculiventris* (Heteroptera: Pentatomidae) on noctuid caterpillars. Bull Entomol Res 96:489–496

Maldonado, J. 1990. Systematic catalogue of the Reduviidae of the world. Caribbean Journal of Science, Special Edition. University of Puerto Rico, Mayagu¨ ez, 694 pp.

Maran, S.P.M. 1999. Chosen reduviid predators- prey interaction, Nutritional and Kairamonal Chemical Ecology. Ph.D Thesis. Manonmaniam Sundaranar University. pp. 111.

Maran, S.P. Babu, M. and Ignacimuthu, A. 2002. Functional response of *Rhynocoris marginatus* (Fab.) (Heteroptera: Reduviidae) on *Spodoptera litura* (Fab.) (Lepidoptera: Noctuidae). Proc. of the Vistas of Entomological Research for the New Millennium. G.S.Gill Research Institute, Guru Nanak College Chennai. (Eds. K.P. Sanjayan, V. Mahalingam and M.C. Muralirangan.). pp.112-115.

Marques, O.M., Hélcio R. Gil-Santana, Maurício Lopes Coutinho and Djael Dias da Silva Júnior. 2006. Percevejos predadores (Hemiptera, Reduviidae, Harpactorinae) em fumo (*Nicotiana tabacum* L.) no município de Cruz das Almas, Bahia. *Revista Brasileira de Zoociências,* 8 (1):55-59.

Maria, S. Gonalez, Jose, Louis Soulages, Rodolf, and Brenner, R. 1991. Changes in the haemolymph lipophorein and very high density lipoprotein levels during the fifth nymphal and adult stages of Triatoma infestans. *Insect Biochemistry,* 21 (6) : 679 - 687.

Mark, A. and Jervis, K. 2005. Insect as natural enemies. A practical perspective published by Springer. 137PP.

Markkula, M., and Roivaines, S. 1961. The effect of temperature, plant food and starvation on the oviposition of some Sitona (Coleoptera: Curculionidae) species. *Annual Entomological Fern,* 27: 30-45.

Matsumara, S. 1988. Studies on the enzyme in the silk worm II. Effect of temperature on the action of enzymes (Consideration on the effect of temperature upon the Silkworm). Bull, Negano, *Sericulture Experimental science,* 6: 1-54.

McIver, J. 1981. An examination of the utility of the precipitin test for evaluation of arthropod predator-prey relationships. *Canadian Entomology,* 113: 213–222.

Mckillip, J.L., Small, C.L., Brown, J.L., Brunner, J.F. and Spence, K.D. 1997. Sporogenous midgut bacteria of the leaf roller, *Pandemis pyrusana* (Lepidoptera: Tortricidae). *Environmental Entomology,* 26: 1475-1481.

Medeiros, R. S., F.S. Ramalho, J. C. Zanuncio, and J. E: Serrao (2003a) Effect of temperature on life table parameters of *Podisus nigrispinus* (Pentatomidae) feb with *Alabama argillacea* (Lepidoptera: Notodontidae) larvae, J. Appl. Ent: 127: 209-213.

Miles, P.W. 1972. The saliva of Hemiptera. *Adv. Ins. Physiol,* 9: 183 – 255.

Miller, N.C.E. 1971. The Biology of Heteroptera (England: E.W.C. Ltd.,) II Edn, 206.

Minja , E . M . , Shanowe r , T . G . , Onguro , J . M . , Nderitu , J . , and Songa, J . M . 1999. Natural enemies associated with arthrpod pests of pigeonpea in Eastern Africa . *Internat ional Chic kpea and Pigeonpea News letter,* 6 : 4 7 - 5 0 .

Mitsuyoshi, T. 2004. Effects of temperature on voiposition in overwintering femals and hatch in first – generation larvae on *Psevdaulacaspis pentagona* (Hemiptera : Diaspididae). *Applied Entomological Zoology,* 3(1); 15-26.

Miyamoto, S. 1961. Comparative morphology of alimentary organs of Heteroptera, with the phylogenetic consideration. *Sieboldia,* 2:197–259.

Morgan, K. R. 1985. Body temperature regulation and terrestrial activity in the ectothermic tiger beetle *Cicindela tranquebarica*. *Ecological Entomology,* 14:419-428.

Moser, D.R., Krichoff, L.V., and Donelson, J.E. 1989. Detection of *Trypanosoma cruzi* by DNA amplification using the polymerase chain reaction. *Jouranl of Insect Microbiology,* 27 : 1477-1482.

Mukerjii, M.K., and Le Roux, E.J. 1965. Laboratory rearing of a Queled strain of the pentatomid predator, *Podisus maculiventris* (Say) (Hemiptera : Pentatomidae*). Phyto- protection,* 46: 40-60.

Murdoch, W.W., Chesson, J., and Chesson, P.L 1985. Biological control in theory and practice. *American Nature,* 135:344-366.

Nagai, K., and Yano, E. 1999. Effects of temperature on the development and reproductive of *Orius slauteri* (Heteroptera: Anthocoridae) a predator of T*hripspalmi Karny* (Thysanoptera : Thripidae). *Applied Entomological Zoology,* 34: 223 -229.

Nakamura, K. 2003. Effect of photoperiod on development and growth in a pentatomid bug. *Dolycoris bacearum. Entomological Science,* 6: 11 - 16.

Navarajanpaul, A.V. 2003. Biological control of lepidopteran pests using important predators. In Biological control of lepidopteran pests. (Eds. P.I. Tandon, C.R. Ballal, S.K. Jalali, R.J. Rabindra). pp.19-22.

Nayar, K.K., Ananthakrishnan, T.N., and David, B.V. 1976. General and Applied Entomology. Tata Mc Grew Hill publishing Co., New Delhi, PP. 169-170.

Neal, J. W., Jr. and Douglass, L. W. 1988. Development, oviposition rate, longevity, and voltinism of Stephanitis pyrioides (Heteroptera: Tingidae), an adventive pest of azalea, at three temperatures. *Environmental Entomology,* 17(5): 827-831.

Neven, L.G. 2000., Physiological response of insects of heat. *Postharvest Biology and Technology* 21: 103 - 111.

Nishi, and Takahasi., L. 2002. Effect of temperature on oviposition and development of *Amphibolus venator* (Klug) (Hemiptera: Reduviidae), a predator of stored product insects, *Jouranal of.Applied Entomological Zoology,* 37 : 415 - 418.

O'Neil, R. J. and Wiedenmann, R. N. 1990. Body weight of *Podisus maculiventris* (Say) under various feeding regimes. *Canadian Entomology,* 122: 285 – 294.

Okasha, A.Y.A. 1968a. Effects of sub-lethal high temperature on an insect, *Rhodnius prodixus* (Stal.) II. Metabolic of cessation and delay of moulting. *Journal of Experimental Biology,* 48: 465-473.

Okasha, A.Y.A. 1968b. Effects of sub-lethal high temperature on an insect, *Rhodnius prolixus* (Stal.) III. Metabolic change and their bearing on the cessation and delay of moulting. *Journal of Experimental Biology,* 48: 475-486.

Okasha, A.Y.K. 1964. Effects of high temperature in *Rhodnius prolixus* (Stal.). *Nature,* 204 : 1221 - 1222.

Omakar, and Pervez, A. 2002. Influences of temperature on age-specific fecundity of a ladybeetle, *Micraspis discolor* (Fab.) *Insect Science Application,* 22(1) : 61 - 65.

Omkar, Shruti Rastogi and Pooja Pandey. 2009. Effect of temperature on reproductive attributes of the Mexican beetle *Zygogramma bicolorata* (Coleoptera: Chrysomelidae). International Journal of Tropical Insect Science 29 (1): 48–52

Oliveira, S.A., Auad, A.M., Souza, B., DANIELA M. SILVA and CAIO A. CARVALHO. 2010. Effect of temperature on the interaction between *Chrysoperla externa* (Neuroptera: Chrysopidae) and *Sipha flava* (Hemiptera: Aphididae). *Eur. J. Entomol.* 107: 183–188

Pankaj, K., Mishra and Tandon, S.M. 2003. Gut bacterial flora of *Helicoverpa armigera* (Hub.) (Lepidoptera : Noctuidae). *Indian Journal of Microbiology,* 43(1): 55-56.

Parajulee, M.N. and Phillips, T.W. 1992. Laboratory rearing and field observations of *Lyctocoris campestris* (Heteroptera: Anthocoridae) a predators of stored product insect. *Annual Entomological Society of America* 85: 736 - 743.

Parajulee, M. N., T. W. Phillips, J. E. Throne, and Nordheim, E. V. 1995. Life history of immature *Lyctocoris campestris* (Hemiptera: Anthocoridae): effects of constant temperatures and relative humidities. *Population Ecology,* 24(4): 889-897.

Prado, S. S., M. Golden, P. A. Follet, M. P. Daugherty, and R. P. P. Almeida. 2009. Demography of gut symbiotic and aposymbiotic *Nezara viridula* (L.) (Hemiptera: Pentatomidae). Environmental Entomology, 38:103–109.

Prado, S. S., Kim Y. Hung, Matthew P. Daugherty, and Rodrigo P. P. Almeida. 2010. Indirect Effects of Temperature on Stink Bug Fitness, via Maintenance of Gut-Associated Symbionts. *Applied and Enviormental Microbiology,* 76 (4): 1261–1266.

Pickel, V.M. 1981. Immunocytochemcial methods. In : Neuroanatomical tract tracing methods. L. Heimer and M.J. Ro Bards (eds). New York; Plenum Press, pp. 483 – 509.

Pigman. W. D., and Horton. D. 1970. The carbohydrates chemistry and biochemistry, 2nd edition, Vol. 2, Academic Press New York.

Pingale, S. V. 1954. Biological control of some stored grain pests by the use of a bug predator, *Amphibolus venator*. Klug. *Indian J. Ent.* 16: 300–302.

Ponnamma, K.N., Kurian, C. and Koya, K.M. 1919. Record of *Rhynocoris fuscipes* (Fabr.) (Heteroptera : Reduviidae) as a predator of Myllocerus curicornis (F.) (Coleoptera: Curcullionidae). *The Agricultural Research Journal of Kerala*, 17: 91-92.

Price, G.M. 1965. Nucleic acids in the larva of the blowfly *Calliphora erythrocephala. Journal of insect Physiology,* 11: 869-878.

Price, P.W. 1984. Insect Ecology, 2nded. Wiley, Newyork, NY 607.

Rabb, R.L. Stinner, R.E. and van den Bosch, R. 1976. Conservation and augmentation of natural enemies. In: (Theory and practice of biological control.) Huffaker, C.B.; Messenger, P.S. (Eds) New York; Academic Press, pp. 233-254.

Ragupathy, E., and Sahayaraj, K. 2002. Biodiversity of reduviid predators in the semi-arid zones of three southern districts of Tamil Nadu. In Proceedings of Vistas of Entomological Research for the New Millenium (Eds. K.P. Sanjayan, V. Mahalingam and M.C. Muralirangan). Pp. 31 – 36.

Reza, M., Otto S., and Keller, M.A. 2008. Factors affecting detachability of prey DNA in the gut contents of invertebrate predators: A polymerase chain reaction based method. *Entomologia Experimentals and Applicata,* 126 : 3 – 194-202.

Rhykman, R.E., and Ryckman, A.E. 1996. Reduviid buys In "Insect customization and mass production" (Ed. Smith, C.N.). Academic press, London, pp 197-199.

Rolf, N., Lesile, A., Willingham, Diane L., Engler, Kenneth J., Tolman, David Bellows, Dick, J. Vander Horst, Gloria, M. and Yepij plasceneia John, H. 1999. A novel lipoprotein from the haemolymph of the cochineal insect, *Dactylophis confuses. Eurpean Journal of Biochemistry*, 261: 285 - 290.

Rosenheim, J.A. 1998. Higher-order predators and the regulation of insect hewrbivore poplulations. *Annual Revieve of Entomology,* 43:421–47.

Rosenheim J.A., Limburg D.D. and Colfer R.G. 1999. Impact of generalist predators on a biological control agent, Chrysoperla carnea: Direct observations. *Ecol. Appl.* 9: 409–417.

Roy, M., J. Brodeur, and Coutier, C. 2002. Relationship between temperature and developmental rate of *Stethorus punctillum* (Coleoptera: Coccinellidae) and its prey *Tetranychus mcdanieli* (Acarina: Tetranychidae). *Environmental Entomology,* 31(1): 177-187.

Ruberson, J. R., and L. H. Williams. 2000. Biological control of *Lygus* spp. A component of area wide management. *Southwestern Entomology* (Suppl. No. 23.) PP. 96-110.

Rudolf, E.J.C., Malausa. P. and Millot Pralavario, R. 1993. Influence of cold temperature on biological characteristics of *Orius laevigatus* and *Orius majunculus* (Het: Anthocoridae) *Entomophaga,* 38 : 317 - 325.

Russomando, G., Figueiredo, A., Almiror, M., Sakamotao M., and Morita K. 1992. Polymerase chain reaction – based detection of *Trypanosoma cruzi* DNA in serum, *Journal of Clinical Microbiology* 30:286-288.

Ryan, R.O., and Dick, J. 2001. Lipid transport Biochemistry and its role in energy production, *Annual Review of Entomology,* 45 :233- 260.

Sahayaraj, K. 1991.   Bioecology, Ecophysiology and Ethology of chosen predatory hemipterans and their potential in biological control (Insecta: Heteroptera: Reduviidae).   Ph. D. Thesis, Madurai Kamaraj University, Madurai, India.

Sackett, T. E., Buddle, C. M. and Vincent, C. 2007. Effects of kaolin on the composition of generalist predator assemblages and parasitism of *Choristoneura rosaceana* (Lep., Tortricidae) in apple orchards. *Journal of applied entomology* 131 (7): 478 – 485.

Sadasivam, S., Manickam, A., 1997. Biochemical methods, second edition. New age international Publication. India. PP 8-9.

Sahayaraj, K. 1994. Capturing success by reduviid predators *Rhynocoris kumarii* and *Rhynocoris marginatus* on different age groups of Spodoptera litura, a polyphagous pest (Heteroptera: Reduviidae). *Journal of Ecobiology,* 6(3): 221 - 224.

Sahayaraj, K. 1995a. Bioefficiacy and prey size suitability of *Rhynocoris marginatus* Fabricius to *Helicoverpa armigera* Hubner of groundnut (Insecta: Heteroptera: Reduviidae); *Fresenius Environmental Bulletin,* 4: 270 – 278.

Sahayaraj, K. 1995b. Developmental stages and biocontrol potential of a reduviid predator, *Acanthaspis pedestris* Stal against termites on groundnut. *International Arachis News letter,* 15: 57-59.

Sahayaraj, K. 1995c. Functional response of the reduviid predator *Ectomocoris tibialis* Distant of the cotton stainer *Dysderus cingulatus* (Fabri.). *Journal of International Study and Research,* 4(2): 65-68.

Sahayaraj, K. 1995d. Bio-efficacy and development of the predator *Rhinocoris marginatus* on *Spodoptera litura. International Arachis Newsletter,* 15: 56-57.

Sahayaraj, K. 1999. Field evaluation of *Rhynocoris marginatus* (Fab.) against two groundnut defoliators. *International Arachis Newsletter,* 19:41 – 42.

Sahayaraj K. 2000. Evaluation of Biological control potential of *Rhynocoris marginatus* on four groundnut pests under laboratory conditions. *International Arachis Newsletter*, 20(1):72-74.

Sahayaraj, K. 2001. Biopesticidal impacts on the biocontrol potential and behaviour of *Rhinocoris marginatus* (Fab.) (Hemiptera: Reduviidae) to groundnut pest *Spodoptera litura* (Fab.). *International Arachis Newsletter*. 21:46 – 48.

Sahayaraj, K. 2002a. Small-scale laboratory rearing of reduviid predator *Rhinocoris marginatus* (Fab.) (Hemiptera: Reduviidae) on *Corcyra cephalonica* Stainton larvae by larval card method. *Journal of Central Europian Agriculture*, 3(2): 137 – 148.

Sahayaraj. K. 2002b. Biopesticide an Indian scenario. *Agribios*, 1 (7): 26-29.

Sahayaraj, K., 2003. Hunter reduviids in cotton bug control. *Agrobios*. 1(12): 9 – 11.

Sahayaraj, K. 2004. Indian Insect Predators in Biological control. (Editor) Dayas Publication, New Delhi. pp. 400.

Sahayaraj, K. 2006. Ecological adaptive featurs of Hunter Reduviids (Heteroptera: Reduviidae: Reduviinae) and their biological contro In: Perspective in animal ecology and reproduction (Volume 3) (Guptha, VK and Verma AK eds.). *Daya Publishing House, New Delhi*. pp 22-49.

Sahayaraj, K. 2007a. Biosafety of Pesticides and Biopesticides. In: Pest control mechanism of Rediviids, Oxford Book company, Narayan Niwas, Jaipur, India. Pp: 106-107.

Sahayaraj, K. 2007b. Isolation, identification and charecterization of gut flora of three reduviid predators. *Asian Journal of Microbiology, Biotechnology and Environmental Science*, 9(4): 1073-1075.

Sahayaraj, K. 2008. Aphid management by predators and Myco-insecticides. *Green Farming*, 1(4): 43-45.

Sahayaraj, K. and Ambrose, D.P. 1992a. Biology, redescription and predatory behaviour of an assassin bug *Allaeocranum quadrisignatum* (Reuter) (Heteroptera: Reduviidae). *Journal of Soil Biology and Eco*logy, 12(2): 120-133.

Sahayaraj, K. and Ambrose, D.P. 1992b. Biology and predatory potential of *Endochus umbrinus* Reuter (Heteroptera: Reduviidae) from *South India*. *Bulletin of Entomology*, 33(1-2): 42-55.

Sahayaraj, K. and Ambrose, D.P. 1993a. Biology and predatory potential of Coranus nodulosus Ambrose & Sahayaraj on *Dysdercus cingulatus* Fabricius and *Oxycarenus hyalinipennis* Costa (Heteroptera: Reduviidae). *Hexapoda*, 5(1): 16-22.

Sahayaraj, K., and Ambrose, D.P. 1993b. Population dynamics of five Reduviids from Reduviids from Kaipothai Scrub Jungles. South India. *Jouranl of Soil Biology And Ecology,* 13: 122-129.

Sahayaraj, K., and Ambrose, D.P. 1994. Functional response of a reduviid predator to two pests. *Biology Education,* 11(2): 114 – 118.

Sahayaraj, K. and Ambrose, D.P. 1996a. Biocontrol potential of the reduviid predator *Neohaematorrhophus therasii* Ambrose and Livingstone (Heteroptera: Reduviidae); *Journal of Advanced Zoology,* 17(1): 49 – 53.

Sahayaraj, K. and Ambrose, D.P. 1996b. Functional response of the reduviid predator *Neohamatorrhophus therasii* Ambrose and Livingstone to the cotton stainer *Dysdercus cingulatus* Fabricius; In Biological and cultural control of insect pests, an Indian scenario, D.P. Ambrose (Ed.) (Tirunelveli, India: Adeline Publishers), pp. 328 – 331.

Sahayaraj K, and Balasubramanian R. 2008. Biological control potential evaluation of artificial and factitious diets reared *Rhynocoris marginatus* (Fab.) on three pest. *Archives of Phytopathology,* 42 (5): 12-18

Sahayaraj, K. and Jeyalakshimi, T. 2002. Mass rearing of *Rhynocoris marginatus* Fab on live and frozen larvae of *Corcyra cephalonica* biology. *Entomologica Croatica.* 6(1-2) : 35-49.

Sahayaraj, K. and Mary J. 2003. Impact of NPV(S) on *Spodoptera litura* (Fabricius) mortality and flora. *Journal of Nature Conservation,* 15(1): 43-50.

Sahayaraj, K., and Martin, P. 2003. Assessment of *Rhynocoris marginatus* (Fab.) (Hemiptera : Reduviidae) as augmented control in groundnut pests. *Journal of Central European Agriculture,* 4(2): 103 - 110.

Sahayaraj, K., and Paulraj, M.G. 1999. Effect of plant products on the eggs of *Rhynocoris marginatus* (Fab.) (Hemiptera : Reduviidae). *Insect Environment,* 5(1) : 23-24.

Sahayaraj, K., and Paulraj, M.G. 2001a. Effect of Cold storage on egg hatching in two reduviid predators *Rhynocoris marginatus* (Fab.) and *R. fuscipes* (Fab.) Hemiptera: Reduviidae). *Beligam Journal of Entomology* .3 : 201-207.

Sahayaraj, K., and Pulraj, M.G. 2001b. Behvaiour of *Rhynocoris marginatus* (Fab.) to chemical cues from three-lepidopteron pest (Heteroptera : Reduviidae). *Journal of Biological control,* 15 (1):1– 4.

Sahayaraj, K. and Paulraj, M.G. 2003. Insect pests and beneficial arthropods of groungnut in relation to wind velocity; *Asian Journal of Microbiology Biotechnology and Environmental Science,* 5(2) : 101 – 103.

Sahayaraj, K. and Raju, G. 2003. Pest and natural enemy complex of groundnut in Tuticorin and Tirunelveli districts of Tamil Nadu. *International Arachis News Letter*. 23: 25 – 29

Sahayaraj, K., and Raju, G. 2006. Assessing the predation of *Rhinocoris marginatus* (Fab.) (Reduviidae) and *Menochilus sexmaculatus* (Fab.) (Coccinelidae) on *Aphis craccivora* (Koch) *Journal Biological Control*, 6:61-67.

Sahayaraj, K. and Selvaraj, P. 2003. Life table characteristics of *Rhynocoris fuscipes* Fab. in relation to sex ratios. *Ecology Environment and Conservation*. 9(2): 115 – 119.

Sahayaraj, K., Nirmala, K. and Selvaraj, P. 2002. Biological control potential of a reduviid predator, *Rhynocoris fuscipes* (Fab.) on three groundnut pests. *Asian Journal of Microbiology, biotechnology and Environmnetal Science*. 4 (4): 451 – 455.

Sahayaraj, K., Martin, P. and Karthikraja, S. 2003. Suitable sex -ratio for the mass rearing of reduviid predator *Rhynocoris marginatus* (Fab.). *Journal of Applied Zoological Research*, 14(1): 34 - 37.

Sahayaraj, K., Thankarani, S., and Delma,J.C.R. 2004. Comparative prey suitability of *Helicoverpa armigera* and *S. litura* larvae for *Rhinocoris marginatus* (Fab.) (Insecta: Heteroptera: Reduviidae); *Belgium Journal of Entomology*, (4) 383-392.

Sahayaraj K, Kumara Sankaralinkam, S. and Balasubramanian R. 2007a. Prey influence on the salivary gland and gut enzymes qualitative profile of *Rhynocoris marginatus* (Fab.) and *Catamiarus brevipennis* (Serville) (Heteroptera: Reduviidae). *Journal of Insect Sciences*, 4(4):331-336.

Sahayaraj K, Venkatesh P, and Balasubramanian R. 2007b. Feeding Behaviour and Biology of a Reduviid Predator *Rhynocoris marginatus* (Fabricius) (Heteroptera: Reduviidae) on Oligidic Diet. *Hexapoda*, 14(1): 24-30.

Sahayaraj, K. Lalitha, C. and R. Balasubramaniam. 2008. Biosafety of *Metarhizium anisopliae* (Metschinikoff) Sorokin on a reduviid predator *Acanthaspis pedestris* (Hemiptera : Reduviidae). *Hexapoda* 15 (1): 46-48.

Salkeld, E.H. 1961. The distribution and identification of esterases in the developing embryo and young nymph of the large milk weed bug *Oncocepettus fasciatus* (Dallas). *Canadian Journal of Zoology*, 39: 589-595.

Salkeld E.H. 1965. Electrophoretic separative and identification of esterases in eggs and young nymphs of the large milk weed bug *Oncocepettus fasciatus* (Dallas), *Canadian Journal of Zoology*, 43: 593-601.

Salt, G. 1965. The cellular Defense Reactions of Insects, Cambridge University Press, Cambridge.

Salt, R.W. 1953. The influence of food on cold handiness of insects. *Canadian entomology*, 85 : 261 - 269.

Santo D, J.W., Hannis, D. and Tiedje, J.M. 1998. Influence of diet on the structure and function of the bacterial hindgut community of crickets. *Microbiol. Ecology*, 7: 761-767.

Santos-Neto, J.R., Mezencio, J.M.S., Chagas, A.T.A., Michereff-Filho, M. and Serrao, J.E.. 2010. Use of Serological Techniques for Determination of *Spodoptera frugiperda* (J E Smith) Predators (Lepidoptera: Noctuidae). *Neotropical Entomology* 39(3):420-423.

Sarah M., Brad Scholz., Sara Armitage, and Johnson. M.L. 2007. Effects of diet, temperature and photoperiod on development and survival of the bigeyed bug. *Geocoris lubra* (Kirkadly) (Hemiptera: Geocoridae). *Bio control,* 1 52 : 63 - 74.

Saxena, K.N., and Bhatnagar, P.L. 1961. Nature and characteristies of invertase in reaction to utilization of scrose in the gut of *Oxycarensus hyalinipennis* (Costa) (Heteroptera : Lygaeidae) *Journal of Insect physiology,* 7: 109-126.

Sayaka, M. T., Imamura., Porintip Visarathanonth and Akihiro M. 2007. Effects of temperature on the development and reproduction of the predatory bug *Joppeicus paradoxus* (Puxton) (Hemiptera : Joppeicidae) reared on *Trifolium confusum* eggs. *Environmental Entomology*, 29 (4: 520-531.)

Schaefer, C.W. 1988. Reduviidae (Hemiptera: Heteroptera) as agents of biological control; In: (Eds., Bicovas, K.S. Ananthasubramanian, P. Venkatesan and S. Sivaraman) Loyola College., *Madras*. 1: 27 – 33.

Schultz, T. D., N. F. Hadley. and Quinlan, M. 1992. Preferred body temperature, metabolic physiology and water balance of adult *Cicindela longilabris*: a comparison of populations from boreal habitats and climatic refugia. *Physiological Zoology,* 65: 226-242.

Shapiro, J.P., Law, J.H. and Wells, M.S. 1988. Lipid transport in insects, *Annual Review of Entomol,* 33 : 297 - 378.

Sheppard, M., McWhorter, R.E. and King, E.W. 1982. Life history and illustrations of *Pristhesancus plagipennis* (Hemiptera: Reduviidae), *Canadian Entomology*, 114: 1089 – 1092.

Sheppard, S.K., Henneman, M.L., Memmott, J. and Symondson, W.O.C. 2004. Infiltration by alien predators into invertebrate food webs in Hawaii: a molecular approach. *Molecular Ecology,* 13: 2077–2088.

Sheppard, S. K. and Harwood, J. D. 2005. Advances in molecular ecology: tracking trophic links through predator–prey food-webs *Functional Ecology*, 19: 751–762.

Sileshi, G., Kenis, M., Ogol, C.K.P.O. and Sithanantham, S. 2001. Predators of *Mesoplatys ochroptera* in sesbania planted-fallows in eastern Zambia. *BioControl,* 46: 289–310.

Shinmizu, T. and Kawasaki, K. 2001. Geographic variability in diapause response of Japanese *Orius sauteri. Journal of Applied Entomological Zoology,* 98: 303 – 316.

Shower, A. T and Greenberg, S. M. 2003. Effects of Weeds on Selected Arthropod Herbivore and Natural Enemy Populations, and on Cotton Growth and Yield. *Environ. Entomol.* 32(1): 39 – 50.

Shrewsburry, P. M. 1996. Factors influenceing the distribution and abundance of the azalea lacebug, *Stephanitis pyrioides* in simple and complex landscapes. Ph.D. Dissertation. University of Maryland at College Park, MD.

Shiankal, Y, M.a., Oschs D,E., Tolezano, J.E., Kirchoff,L.V. 1996. Use of PCR for detecting *Trypanosoma cruzi* in Triatomine vectors. Trans R. So *Tropical Medical Hygyne,* 90: 649-651.

Silva, I.G., 1985. Influencia da temperatura na biologia de triatomineos. I. *Triatoma rubrovaria* (Blanchard, 1843) (Hemiptera, Reduviidae). *Revista Goiana de Medicina,* 31 : 1 –37.

Singh, O.P. 1985. New record of *Rhynocoris fuscipes* (Fabr.) as a predator of *Dicladispa armigera* (Oilver): *Agricultural Science Digest,* 5 (3): 179-180.

Singh, O.P., and Gangrade, G.A. 1975. Parasites, Predator and diseases at larvae of *Diacrisia deligna* walker (Lepidoptera : Arctidae) on soybean. *Current Science,* 44 : 481-482.

Singh, O.P., and Sing, K.J. 1987. Record of *Rhynocoris fuscipes* Fabricus as a predator of Green stink bug: *Nezara viridula* Linn. Infesting soybean in India. *Journal of Biological Control,* 1: 143-146.

Sitaramiah, S., and Satyanarayana, S.V.V. 1976. Biology of *Harpactor costalis* stal. (Heteroptera: Reduviidae) on tobacco caterpillar *Spodoptera litura. Frontier of Tropical Research,* 2: 134 – 136.

Smith, D.C. and Douglas, A.E. 1987. The biology of symbiosis. London: Arnold. pp. 302.

Sopp, P. I., and Sunderland, K. D. 1989. Some factors affecting the detection period of aphid remains in predators using ELISA. *Entomological Experiment Application,* 51: 11–20.

Sopp, P. I., Sunderland, K. D., Fenlon, J. S., and Wratten, S. D. 1992. An improved quantitative method for estimating invertebrate predation in the field using ELISA. *Journal of Applied Ecology,* 29: 295–302.

Stamp, N. E., Yang, Y. and Osier, T. 1997. Response of an insect predator to prey fed multiple allelochemicals under representative thermal regimes. *Ecology,* 78(1): 203-214.

Stern, V.M., Smith, R.F. Vandan Bosch, R., and Heagen, S. 1990.The integrated control concept. *Hillgardia,* 29: 81- 89.

Stuart, M. K., and Greenstone, M. H. 1990. Beyond ELISA—A rapid, sensitive, specific immunodot assay for identification of predator stomach contents. *Annual Entomological Society of America*, 83: 1101–1107.

Sundarland, K.D., N.E., Stacey, D.L., Fuller, B.J. 1978. A study of Aphid feeding by polypagous predators on cereal aphids using ELISA and gut dissection. *Journal of Applied Entomology,* 24: 970-9

Sunderland, K.D. 1988 Quantitative methods for detecting invertebrate predation occurring in the field. *Annals of Applied Biology,* 112: 201–224.

Sunderland, K.D. 1996. Progress in quantifying predation using antibody techniques. The Ecology of Agricultural Pests: Biochemical Approaches (Eds W.O.C. Symondson and Liddel, pp:414-455.

Sunderland, K.D., Crook, N.E., Stalay D.L., and Fuller, B.J. 1987. A study of aphid feeding by polyphagous predators on cereal aphids using ELISA and gut dissection. *Journal of Applied Ecology,* 24 : 907-933.

Symondson, W. O. C., and Liddell, J. E. 1993. Differential antigen decay rates during digestion of Molluscan prey by carabid predators. *Entomological Experimental Application,* 69: 277–287.

Symondson, W.O.C. 2002. Molecular identification of prey in predator diets. *Molecular Ecology,* 11: 627–641.

Tanada, Y. and Kaya, H.K. 1993. Insect Pathology, *Academic Press, London.* pp. 276.

Tawfik, M.F.S. and Awadallah, T. 1983. Effect of temperature and relative humidity on various stages of the predated, *Allaceocramm biannulipes* (Montr.et. Sign) (Hemiptera : reduviidae). *Bulletin of society of entomology,* 64:239-250.

Tawfik, M.F.S., Awadallah, K.T. and Abdellah, M.M.H. 1983a. The biology of the reduviid *Allaeocranum biannulipes* (Montrouzier and Signoret), a predator of stored product insects; *Bulletin of Social Entomology,* 64: 231 – 237.

Tawfik, M.F.S., Awadallah, K.T. and Abdellah, M.M.H. 1983b. Effect of prey on various stages of the predator, *Alloeocranum biannulipes* (Montr. et Sig.) (Hemiptera: Reduvidae); *Bulletin of Social Entomology Egypt,* 64: 251 – 258.

Tefler, W.H., Kain, P.S., and Law J.H. 1983. Arylphorin a new protein from *Hyalophora cecropia* comprisons with calliphorin and manducin, *Insect Biochemistry,* 13: 601 – 613.

Teresa, T., Shirley, B., and Eilence, M.L. 2002. Biotechnology : DNA to Protein. *A Laboratory project in Molecular Biology. Tata Mc Graw Hill.* 81PP.

Thompson, D. J. 1978. Towards a realistic predator–prey model: the effect of temperature on the functional response and life history of larvae of the damselfly Ischnura elegans. *Journal of Animal Ecology,* 47: 757–767.

Thomashow, M.F., 1998. Update on adaptation to physical stress: role of coldresponsive genes in plant freezing tolerance. Plant Physiology 118, 1–7.

Torres. J.B., Zanuncio, J.C., and De oliveira, M.N., 1998. Nmphal development and adult reproduction of the stinkbug predator *Podisus nigrispinus* (Heteroptera : Pentatomidae) under fluctuating temperatures. *Journal of Applied Entmology,* 122: 509-514.

Tsuchida, T., Koga, R., Shibao, H., Matsumoto, T. and Fukatsu, T. 2002. Diversity and geographic distribution of secondary endosymbiotic bacteria in natural populations of the pea aphid, *Acyrthosiphon pisum. Molecular Ecology,* 11: 2123–35.

Tustomu, T., Ryuichi, K., Makiko, S., and Takema, F. 2006. Facultative bacterial endosymbionts of three aphid species, *Aphis craccivora, Megoura crassicauda* and *Acyrthosiphon pisum,* sympatrically found on the same host plants. *Applied Entomological Zoology,* 41(1): 129-137.

Upadhyay, V.B., and Misra, A.B. 1991. Nutritional ability of bivottine silkworm. *Bombyz mori* L. larvae at higher temperature regimes. *Journal of Advanced zoology,* 12 (1) : 56 - 59.

Upadhyay, V.B., and Misra, A.B. 1994. Influence of temperature on the passage of food through the gut of multivoltine *Bombyx mori* (L.) Larvae, *Indian Journal of Sericulture,* 33(2) : 183 - 185.

Usharani, P. 1992. Temperature- Induced Effects on Predation and growth of *Eocanthecona furcellata* (wolf) (Pentatomidae: Heteroptera). *Journal of Biological. Control,* 6 (2) : 72 – 76.

Uvarov, B. 1996. Grasshoppers and locusts. A handbook of general acridology. Vol. 1. Alimentary system. Cambridge Unversity press. pp. 79-89.

Vallejo, G.A., Guhl, F., Chiari, E., and Macedo, A.M. 1999. Species specific detection of *Trypanosoma cruzi* and *Trypanosoma rangeli* in vector and mammalian hosts by polymerase chain reaction amplification of kinetoplasts minicirele DNA. *Acta Tropical Entomology,* 72 : 203 – 212.

Van Weeman, B.K. and Schuurs, A.H.W.M. 1971. Immunoassay using antigen enzyme conjugates. *FEVS Lett*er, 15 : 232 - 236.

Vennison, S. J. 1988. Bioecology and Ethology of assassin bugs (Insecta: Heteroptera: Reduviidae). Ph. D. Thesis, Madurai Kamaraj University, Madurai, India.

Vennison, S.J., and Ambrose, D.P. 1988. Impact of space of on stadial period, adult longevity, morphometry, oviposition, hatching and prey capturing in *Rhycocoris marginatus* fabricus (Insecta : Heteroptera : Reduviidae). *Jornal of Mittle Zoology Musium Berlin,* 64: 3249 – 355.

Vennison, S.J., and Ambrose, D. P., 1990. Egg development in relation to soil moisture in two species of reduviids (Heteroptera : Reduviidae). *Journal of Soil Biology and Ecology*, 10(2): 116-118.

Weber, D.C., Lundgren, J.G., 2009a. Assessing the trophic ecology of the Coccinellidae:Their roles as predators and prey. Biological Control 51, 199–214.

Weber, D.C., Lundgren, J.G., 2009b. Quantification of predation using qPCR: effect of prey quantity elapsed time, chaser diet, and sample preservation. Journal of Insect Science 9, 41.

Westich, R. and Judith Hough-Goldstein (2001) Temperature and host plant effects on predatory stink bugs for augmentative biological control. Biological Control, 21 (2): 160-167

Whitteman D. 2005. Insect phenotypic Plasticity: Diveristy of response editor: T.N. Ananthakrishnan, 1-57808-322-2, p-210.

Wigglesworth, V.B. 1972. The principles of insect physiology. *Chapman and Hall, Ltd., London*, 827.

Wigglesworth, V.B. 1974. The principles of insect physiology. *Chapman and Hall, Ltd., London*, 827. Whiteman D. 2005. Insect phenotypic Plasticity: Diveristy of response editor: T.N. Ananthakrishnan, 1-57808-322-2, p-210.

Wignall, A.E. and Taylor, P.W. 2008. Biology and life history of the araneophagic assassin bug *Stenolemus bituberus* including a morphometric analysis of the instars (Heteroptera : Reduviidae). *Journal of Natural History*, 42: 59 – 76.

Wignall, A.E. and Taylor, P.W. 2009. Responses of an araneophagic assassin bug, *Stenolemus bituberus* , to spider draglines. *Ecological Entomology*, 34, 415–420.

Williams, J.G.K., Kubelick, A.R., Livak, J., Rafaskai, J.V., Tingey, S. V. 1990. DNA Polymorphism amplified by arbitrary primers are useful as genetic markers. *NHC.Acid. Res.*, 18: 6531-6535

Woolfolk, S.W., Cohen, A.C., Inglis, G.D., 2004. Morphology of the alimentary canal of Chrysoperla rufilabris (Neuroptera: Chrysopidae) adults in relation to microbial symbionts. Annals of the Entomological Society of America 97, 796–808.

Wyatt, G.R. 1976. The biochemistry of sugars and polysuharides in insects. *Adv. Insect Physol,* 4 : 287 – 360.

Zachaniarren, K. E. 1985. Physiology of cold tolerance in insects. *Physiological Review,* 65: 799 - 832.

Zhang, G.F., Chuang, Z., Wan, F.H., and Levei, G.L. 2007. Real – time PCR quantification of *Bemisia tabacci* (Homoptera : Aleyrodidae) B. biotype remains in predators gut : *Molecular Ecology,* 7 (6) : 947– 954.

Zulkefli, M., Norman, K. and Basri, M W. 2004. Life cycle of *Sycanus dichotomus* (Hemiptera: Pentatomidae) – a common predator of bagworm in oil palm. *Journal of oil palm research* 16 (2): 50 – 56.

INDEX

E

F

G

M

N

S

T